"*Why Transportation Fails* offers a much-needed and timely analysis of the multiple social and spatial injustices characterizing mobility systems and practices in South Africa's cities. By synthesizing thinking across disciplinary and theoretical traditions, the book begins to open up new ways of imagining more inclusive and just mobility futures in Africa and beyond. An essential read for social scientists, engineers and planners looking for thinking at urban issues in and from the Majority World."

Tim Schwanen, *Professor of Transport Geography and Director of the Transport Studies Unit, University of Oxford*

"A captivating and insightful book that looks beyond standard tropes about state failure in Africa and instead asks *why* and *how* transportation in South Africa (and elsewhere) could be effective and just. Essential reading for anyone interested not just in how to navigate cities, but how to re-frame urban 'problems' from new perspectives."

Charlotte Lemanski, *Professor of Urban Geography, University of Cambridge*

"*Why Transportation Fails* is a cutting-edge exploration of urban disappointment. Launched with optimism, many transportation initiatives in post-apartheid South Africa have been let-downs. However, this analysis moves beyond negativity to show how using new frames of analysis provide the resources for rethinking urban transformation."

Philip Harrison, *South African Research Chair in Spatial Analysis and City Planning, University of the Witwatersrand*

"Transport geography is always in need of texts that not only critique existing praxis but also offer fresh conceptual and practical insights. *Why Transportation Fails* seeks to reimagine what successful transport systems can be by blending bold intellectual ambition with radical approaches, including animal geographies, decoloniality, and art geographies. Building on these approaches, Astrid Wood offers an empirically rich exploration of South Africa's transport system that critically addresses the entanglements of human-environment relations and urban mobilities. This enables *Why Transportation Fails* to challenge conventional narratives of failure while offering alternative approaches to transport and urban transformation, making it valuable reading for transport and urban planners, policymakers, and scholars across multiple disciplines."

James Esson, *Professor of Geography, Queen Mary University of London*

Why Transportation Fails

Why Transportation Fails offers an in-depth critique of transportation failures in South Africa and beyond. Minibus taxis are unreliable and overcrowded; buses and trains are old and poorly maintained; and new services are derailed by spiraling capital and operational costs, protracted and politicized rollouts, and unsolvable infrastructural challenges. Rather than focusing on these failures, the conceptual and practical analyses in this book develop a critical understanding of transportation and highlight the opportunities for transportation to be more inclusive, innovative, and sustainable.

This book applies cutting-edge scholarship from geography and urban studies – decolonial geographies, Black geographies, policy mobilities, walking geographies, animal geographies, art geographies, and health geographies – to reframe transportation failure. The interdisciplinarity of this scholarship builds an approach that not only tells the story of transportation in South Africa but also uncovers a range of conceptual interpretations and imaginations that stretch beyond urban mobilities and urban development. In decentering traditional geographies of knowledge production, *Why Transportation Fails* contributes to critical considerations of urban transportation in Africa and aspects of transportation justice within social and spatial transformation.

Given the never-ending financial and political investment in solving transportation, *Why Transportation Fails* is an essential reading for scholars of architecture, development, geography, politics, sociology, and urban planners and practitioners.

Dr Astrid Wood is an Associate Professor in the School of Geography, Politics, and Sociology at Newcastle University in the United Kingdom. Astrid is an urban geographer specializing in governance, infrastructure, and transportation. She is the author of *How Cities Learn: Tracing Bus Rapid Transit in South Africa* (2022) as well as over 40 peer-reviewed book chapters and journal articles. She is an editor at *Urban Studies* and at the *Journal of Transport Geography*.

Questioning Cities

The "Questioning Cities" series brings together an unusual mix of urban scholars under the title. Rather than taking a broadly economic approach, planning approach, or more sociocultural approach, it aims to include titles from a multidisciplinary field of those interested in critical urban analysis. The series thus includes authors who draw on contemporary social, urban, and critical theory to explore different aspects of the city. It is not therefore a series made up of books that are largely case studies of different cities and predominantly descriptive. It seeks instead to extend current debates through, in most cases, excellent empirical work, and to develop sophisticated understandings of the city from a number of disciplines including geography, sociology, politics, planning, cultural studies, philosophy, and literature. The series also aims to be thoroughly international where possible, to be innovative, to surprise, and to challenge received wisdom in urban studies. Overall it will encourage a multidisciplinary and international dialogue always bearing in mind that a simple description or an empirical observation that is not located within a broader theoretical framework would not – for this series at least – be enough.

Urban Political Ecology in the Anthropo-Obscene
Interruptions and Possibilities
Edited by Henrik Ernstson and Erik Swyngedouw

Between Bohemia and Suburbia
Boburbia in the USA
William J. Weston

Why Transportation Fails
Critiques from South Africa and Beyond
Astrid Wood

For more information about this series, please visit: www.routledge.com/Questioning-Cities/book-series/SE0756

Why Transportation Fails

Critiques from South Africa and Beyond

Astrid Wood

Routledge
Taylor & Francis Group
LONDON AND NEW YORK

First published 2026
by Routledge
4 Park Square, Milton Park, Abingdon, Oxon OX14 4RN

and by Routledge
605 Third Avenue, New York, NY 10158

Routledge is an imprint of the Taylor & Francis Group, an informa business

British Library Cataloguing-in-Publication Data
A catalogue record for this book is available from the British Library

ISBN: 978-0-367-33840-4 (hbk)
ISBN: 978-1-041-08242-2 (pbk)
ISBN: 978-0-429-32228-0 (ebk)

DOI: 10.4324/9780429322280

Typeset in Times New Roman
by codeMantra

Contents

List of figures *xi*
List of abbreviations *xiii*
Acknowledgments *xv*
Preface *xvii*

1 Introduction 1
Starting with Failure 1
Transportation in South Africa and Beyond 3
Conceptual Contributions 10
Methodologies of Failure 15
Navigating This Book 18

2 Decolonizing Transportation 21
Introduction 21
Decolonizing Geography 22
Decolonizing Geography in South Africa 25
Bringing Decolonizing Discourses into Transportation 27
Decolonizing Transportation in South Africa 30
Conclusion 37

3 Black Transportation 38
Introduction 38
The Whiteness of Geography 39
Defining Black Geographies 41
Bringing Black Geographies into Transportation 44
Black Transportation in Johannesburg 47
Conclusion 51

4 People-Oriented Transportation 53
Introduction 53
Problematizing Transit-Oriented Development 54
Bringing Policy Mobilities into Transportation 57
People-Oriented Transportation in Cape Town and Johannesburg 58
A South African Approach to Transportation Planning 63
Conclusion 65

5 Walking-As-Transportation 67
Introduction 67
Reconceptualizing Walkability 68
Walkability in African Cities 71
Walking-As-Transportation in Johannesburg 74
The Failings of Walkability in Johannesburg 76
Conclusion 82

6 Trans-paw-tation 84
Introduction 84
Finding the Animals in Human Geography 85
An Urban Paws 87
Bringing Animal Geographies into Transportation 89
Trans-paw-tation in South African Cities 92
Conclusion 96

7 Transp-art-ation 98
Introduction 98
Art in Postapartheid South Africa 100
Bringing Art into Transportation 103
Transp-art-ation in South Africa 106
Conclusion 114

8 When Transportation Fails 115
Introduction 115
Public Health, Epidemiology, and Urbanization 116
Disease and the South African City 118
Transportation and COVID-19 119
Impact of COVID-19 on Transportation in South Africa 122
Conclusion 126

9 Conclusion 127
Introduction 127
Reflecting on Why Transportation Fails 127
Reflecting on Transportation in South Africa... 130
Reflecting on Transportation ... Beyond 132

Appendix A *135*
References *139*
Index *169*

Figures

	Minibus taxis parked at the Metro Mall Taxi Rank in Johannesburg (photograph by author, March 2014)	xviii
1.1	Shosholoza Meyl Railways in Johannesburg	4
1.2	Gautrain Sandton Station	5
1.3	Metrobus in Johannesburg's central business district	6
1.4	Rea Vaya Station in Johannesburg	7
1.5	Non-motorized transportation in Gauteng	8
2.1	Seats on Rea Vaya bus	32
2.2	Cycling in Johannesburg	34
2.3	Minibus taxi in Johannesburg	35
4.1	Klipfontein Road in Gugulethu, Cape Town	60
4.2	New developments along Louis Botha Avenue in Johannesburg	62
4.3	No trading zone at Metrobus bus stop in Johannesburg	64
5.1	Pedestrians walk in blocked cycle lanes in Hillbrow, Johannesburg	75
5.2	Map of Johannesburg's Corridors of Freedom and pedestrian bridges	78
5.3	The Grayston Pedestrian and Cycle Bridge	79
7.1	Mural of Rea Vaya bus on Louis Botha Avenue in Johannesburg	106
7.2	Glass Etching on Rea Vaya station in Johannesburg	108
7.3	Art on MyCiTi station wall in Cape Town	109
7.4	Art at the Bree Street Taxi Rank in Johannesburg	110

Abbreviations

AAG	Association of American Geographers
ANC	African National Congress
BEN	Bicycle Empowerment Network
BID	Business Improvement District
BME	Black and Minority Ethnic
BRT	Bus Rapid Transit
DA	Democratic Alliance
EFF	Economic Freedom Front
FIFA	Fédération Internationale de Football Association
GABS	Golden Arrow Bus Services
GDS	Growth and Development Strategy
ITDP	Institute for Transportation and Development Policy
ITP	Integrated Transport Plan
JDA	Johannesburg Development Authority
MK	uMkhonto we Sizwe
MMC	Member of Mayoral Committee
NMT	Non-Motorized Transportation
NTA	National Taxi Alliance
PETA	People for the Ethical Treatment of Animals
POT	People-Oriented Transportation
PRASA	Passenger Rail Agency of South Africa
PRT	Personal Rapid Transit
RGS-IBG	Royal Geographical Society-Institute of British Geographers
PUTCO	Public Utility Transport Corporation
SACN	South African Cities Network
SANRAL	South African Roads Association Ltd.
SANTACO	South African National Taxi Council
SATC	Southern African Transport Conference
TRF	Taxi Relief Fund
TRP	Taxi Recapitalization Program
TOD	Transit-Oriented Development
UCT	University of Cape Town
UNEP	United Nations Environment Program

UN-Habitat	United Nations Human Settlements Program
UK	United Kingdom
US	United States
WHO	World Health Organization

Acknowledgments

This book is the product of many years of research and it would be impossible to thank all of those who have supported it.

To my friends and colleagues in South Africa and those who gave me so many hours of their time, I hope this work provides inspiration to continue improving urban life for all residents.

I am thankful to my friends and colleagues in Newcastle and around the world who have offered feedback on book chapters, conference presentations, grant proposals, and journal articles as well as direction for how to navigate the academic netherworld.

I am lucky to have a supportive husband and two lovely little ones to brighten even the cloudiest of days, and an ever-widening circle of family and friends to keep me motivated through all the starts and stops of writing a book.

I dedicate this work to my late father whose passion for social justice, in spite of apartheid, imbued me with a similar commitment to community engagement. His dedication to transformation, along with so many others, has helped make South Africa a better place.

This book is for all those who still believe.

Preface

I'm reading *Last Stop on Market Street* by Matt de la Peña with my children for the umpteenth time. I can read this book in the dark, without looking at any of the words or pictures, and I know because I read it almost every night. This book tells a sweet story about CJ and Nana riding the bus to volunteer at the soup kitchen – there's a bus that breathes fire, a lady with butterflies in a jar, the driver Mr. Dennis always has a trick for CJ and a busking guitarist lifts the passengers, even the teenagers listening to their iPod, out of the bus and into a magical world. If only transportation was really like this. Instead, I stammer and stumble over the words as I explain to my children that we are going to miss an activity because the bus is late, that the bus isn't running because drivers are forced to go on strike, or that we can't transfer buses as many times as we want because our ticket doesn't work like that. These disappointments don't exist in children's books but they are an everyday part of public transportation in every city around the world. How do we make sense of these failings? Why does transportation fail? This book is an attempt to understand the failures of public transportation. It starts from my own experiences trying to understand the failures of public transportation in urban South Africa but moves beyond the specifics of South Africa to think about transportation through concepts in animal geographies, art geographies, Black studies, decolonial studies, and health geographies. It does not try to solve the technical failings of public transportation – I leave that to next brave researcher – but in understanding why transportation fails, I hope that it opens the door to better solutions, ones that more closely resemble those in our children's books and fairy tales.

Let's go back a bit further: it's February 2020 and I'm finishing the last part of my fieldwork before my sabbatical ends, before my children are born, and before the world shuts down to COVID-19. I am sitting in another meeting with a friendly group of transportation engineers in Johannesburg who have graciously agreed to meet with me, give me their time, and share their experiences working to improve public transportation. In responding to one of my questions, they turn the question to me. They ask, what has happened with Rea Vaya and what should we do to fix it? They ask me – a researcher living in the United Kingdom who admittedly has written the book on Rea Vaya but who is hardly an expert on transportation engineering. And while this is not the first time that an interviewee has turned my questions around on me, I have yet to come up with a clever response.

Instead, I write a big, bold question mark in my notebook. When I review my notes from this research visit, I notice a lot of these same big, bold question marks from other interviews. It seems that the officials responsible for designing, building, operating, and maintaining public transportation are at a loss. And this is definitely not meant to be a critique of them – we certainly don't need another critique of transportation in Africa! – but rather this book comes from my attempt to reconcile my involvement with public transportation transformation in South Africa and the reality that nobody has any idea what to do.

By the time, this book goes to print, even I have to admit that South Africa's noble experiment with bus rapid transit has failed. What began as an exploration of the process through which transportation was leading to urban transformation morphed into a disappointing study of failure. This book is my attempt to understand why Rea Vaya buses don't breathe fire, there are no butterflies or Mr. Dennis on the buses either. Instead, public transportation – especially bus transportation – is populated by downtrodden commuters, retirees, and the occasional do-gooder. This book is not a fairy tale for children but instead tries to understand why transportation fails in the hope that one day it won't fail anymore.

Minibus taxis parked at the Metro Mall Taxi Rank in Johannesburg (photograph by author, March 2014)

1 Introduction

Starting with Failure

Urban transportation in South Africa and beyond is more than a system by which people move from A to B; it is a portal between urban and rural, White and Black, rich and poor, resident and migrant, employed and economically disadvantaged, and so on, all of which have blurred and solidified as cities sprawl. Transportation is central to generating economic and educational opportunities by enabling the movements of people and goods; it provides places of sociality where people come together and develop shared experiences and can also produce spaces of political contestation through which people support and/or oppose the state or the lack thereof. But transportation in South Africa is failing – both government and the private sectors have tried without success to develop efficient or at least not inefficient, popular or at least not reviled, and profitable or at least economically viable transportation systems. These initiatives have increased exponentially in the past two decades with little effect. Around the world, and particularly in South African cities, transportation is failing. This book attempts to understand why.

This story of failing transportation focuses on South Africa, where the majority of the population relies on public transportation. About 10 million of South Africa's 14 million households use public transportation as their main mode of travel, while just 3 million use private transportation (National Department of Transport 2013). Of total commutes, 65 percent of trips are made by taxi (e.g., minibus taxis) with the remaining 20 percent by bus (e.g., private, subsidized buses, and municipal buses), and 15 percent by train (e.g., Metrorail and Gautrain) (Mndebele 2019). The metropolitan municipalities in Cape Town, eThekwini, Johannesburg, and Tshwane are home to extensive bus systems (e.g., Golden Arrow, PUTCO, and People Mover), bus rapid transit (BRT) (e.g., MyCiTi, Go Durban, and Rea Vaya), and rail (e.g., Metrobus in Johannesburg); while a semiformal, under-regulated system of minibus taxis moves the majority of commuters; and across the country, a small proportion of the populace – 300,000 households – use non-motorized transportation as their main mode (South Africa Competition Commission 2021). The economic, political, and social importance of public transportation is therefore tremendous – it is not just about how people get around but also a fundamental aspect of South African life. And yet, despite three decades of programs and

DOI: 10.4324/9780429322280-1

policies, transportation remains shadowed by legacies of apartheid spatial planning and postapartheid neoliberalism. In understanding why transportation fails, this book highlights a complex web of formal and informal, motorized and non-motorized, and modern and obsolete transportation infrastructures and systems.

This book grapples with the concept of transportation failure. It does not focus exclusively on the traditional interpretations of failure, such as the failures of transportation maintenance – and how roads and railways deteriorate over time, transportation infrastructure – and how transportation systems are characterized by inefficient routes, inconvenient stops, and a lack of integration between different modes of transportation, failures of funds – and how costly investments are inadequately financed by bankrupt governments – or emergencies – nor the outcomes of accidents and disasters. Instead, it moves beyond these technical understandings of failure to reconsider the concept by drawing on a broader set of geography literature to examine failure and critically analyze how transportation policies and practices are shaped and styled into ostensible successes.

Geographers have engaged with these questions through a range of scholarship. Ward (2006, 70) explains that "There is nothing natural about which policies are constructed as succeeding and those that are regarding as having failed;" while Temenos and Lauermann (2020) reason that failure is relational (McCann and Ward 2015), multi-temporal (Wood 2015a,b), and only one of many possible outcomes (Wells 2014). For Chang (2017), "failure matters" as it remains both mobile and influential, and Wells (2014, 475) understands "policy failing" as the "interruptions, exceptions, and stalled attempts at policymaking". Lovell (2017) therefore suggests that failure scholarship needs to engage with understandings of absences and mutations as well as with the sites of failure.

Research of policy failure is especially attuned to the ways in which failure is procreative with outcomes greater than or different from those intended. Elsewhere, I consider the circulation of both incomplete and misconstrued understandings of town planning as well as the implications of that failure on both the policy actors and the locality wherein failure takes place (Wood 2019a). Here, I argue that, across colonial and postcolonial African cities, there are instances of introspective failure in which cities extracted planning ideas from colonial cities like London, as well as extrospective transfer in which cities deployed transfer agents to search for ideas elsewhere and then introduce them locally. Moreover, Lauermann (2016) suggests that even failed Olympics bids fuel investment long after the process has ended. These studies are less concerned with the failure to make policy – as might be measured by a community or council opposition – and more on how policy fails once it is approved – as might be measured by failures to reach project outcomes – as well as the unintended results of that failure. This book advances these conceptual discussions by considering the failures of transportation in South Africa and beyond.

These conceptual considerations connect transportation literature with contemporary conversations in geography, politics, and sociology, in addition to urban planning and urban studies. It draws implicitly on, and reinforces, two decades of the mobilities paradigm (Adey et al. 2014; Cresswell 2014; Hannam, Sheller, and Urry 2006; Sheller and Urry 2016). Within these discussions, this book makes two specific contributions to urban transportation in Africa: first, it contributes to critical conversations about urban mobilities in Africa; and second, it develops a richer understanding of transportation justice. To achieve these intellectual aims, this book draws on lively debates in decolonial geographies (Schwanen 2018a; Wood, Kębłowski, and Tuvikene 2020), Black geographies (Alderman, Williams, and Bottone 2022; Inwood, Alderman, and Williams 2015), policy mobilities (McCann and Ward 2011; Wood 2022a), walking geographies (Behrens 2005; Zavetoski and Agyeman 2015), animal geographies (Wolch, West, and Gaines 1995; Wolch 2002), art geographies (Bissell and Fuller 2017; Hawkins 2019), and health geographies (Ali and Keil 2007, 2008; Wolf 2016). The intellectually diverse scholarship helps to develop a pluralistic understanding of transportation and demonstrates a need for creative and experimental approaches to transportation, far beyond the normative, technocratic discussions. In so doing, it not only tells the story of transportation in South Africa but also uncovers a range of conceptual interpretations and imaginations of transportation within an urbanized landscape.

The chapter now proceeds by grounding this study of transportation within South African cities. It then outlines the main conceptual contributions of this book and the overarching arguments that are explored throughout. Next, it discusses my research methodologies as well as my positionality as a researcher. Finally, it concludes with the chapter summaries.

Transportation in South Africa and Beyond

Governmental responsibilities for transportation in South Africa are difficult to untangle. In 1994, public transportation responsibilities were split among the national government maintaining responsibility for rail, the provincial government managing buses, and cities responsible for transportation planning. The National Department of Transport became an advisor with several quasi-autonomous national enterprises handling the technical decisions – for example, the Passenger Rail Agency of South Africa (PRASA), the state-owned enterprise responsible for passenger rail services, and the South African National Roads Authority (SANRAL), the state-owned enterprise responsible for South Africa's road network. Supporters of this approach argued that the National Department of Transport could be more effective in its devolved state, as "a regulator of bureaucratic detail, a provider of infrastructure, and a transport operator" with less focus on policy formulation and strategic planning (National Department of Transport 1996). Those opposed to this approach reason that this positioned the national government, and to a lesser extent the provincial government, as a "policy maker" while local government became the "policy taker" (van Ryneveld 2006, 16). This complex bureaucracy results in divisions not only between national and local departments but also between government and populace. It also leads to duplicate services even within low-density, low-populated areas, and high operating subsidies to areas with little off-peak demand [28].[1]

Rail is the most comprehensive public transportation service in South Africa (see Figure 1.1). The government owns and operates an extensive urban rail

Figure 1.1 Shosholoza Meyl Railways in Johannesburg (photograph by author, March 2014). Shoshololoza Meyl, the passenger rail service operating between South African cities, carries about four million passengers annually.

network – before March 2020, Metrorail, South Africa's rail transportation service, moved about two million people a day in Gauteng, KwaZulu-Natal, the Eastern Cape, and the Western Cape (Nkosi 2020). Trains are preferred over buses or taxis when they are the cheaper option [5]. However, the services are unreliable, trains are often overcrowded, the rolling stock is old and poorly maintained, and there is little security despite high rates of fare evasion, cable theft, and other criminal activity (South Africa Competition Commission 2021). And when the trains stopped running during the COVID-19 pandemic, 80 percent of the railway's stations were looted, making operations impossible to resume. Rail, however, is considered a "social service" and is heavily subsidized – it draws 51 percent of the national transportation budget but moves just 9 percent of passengers (Nkosi 2020).

Another rail service, the Gautrain, an 80-kilometer rapid railway connecting Johannesburg with Tshwane and OR Tambo International Airport, has become an example of "what not to do" in transportation, first because of its high price tag and lengthy construction process and second because of its low ridership and high operating costs [10] (see Figure 1.2). At a cost of R30 billion, the Gautrain promised to reduce traffic congestion and encourage sustainable transportation, stimulate economic growth by creating jobs in construction and train operations, and promote tourism (Van der Westhuizen 2008). While construction only started in February 2006, planning began as early as 1997, when then-Premier of Gauteng Province

Figure 1.2 Gautrain Sandton Station (photograph by author, April 2014). The Gautrain, which connects Johannesburg with Tshwane and OR Tambo International Airport, opened in June 2010 at a cost of R30 billion.

Tokyo Sexwale explored the possibility of introducing the first high-speed inter-urban transportation system on the African continent; and, in 2000, then-Premier Mbhazima Shilowa broadcast his approval for the train. It took another two years however to conduct the environmental impact assessment for the project, and it was not until December 2005 that the project received final approval (Thomas 2013). This extensive deliberation on such a costly project is not surprising, nor was it a coincidence that the initial track to bring spectators to/from the airport became operational in June 2010, just before the 2010 FIFA World Cup (Wood 2017).

The story of transportation failure repeats for buses: there are 19,000 public transportation buses operating in South Africa, which move 912 million passengers per annum. The industry directly employs 34,200 people with another 171,000 indirectly employed nationwide. South Africa's largest private provider is PUTCO (Public Utility Transport Corporation), which operates a fleet of 1,300 buses along subsidized routes in Gauteng and Mpumalanga; with Golden Arrow Bus Services (GABS) as the second largest provider, operating 1,100 buses along more than 2,000 routes in the Western Cape (South Africa Competition Commission 2021). City services are operated in eThekwini (People Mover), Johannesburg (Metrobus), and Tshwane (Tshwane Bus Service) (see Figure 1.3). During apartheid, South Africa's urban bus industry was used as a policy instrument to support uneven development. The bus system was intrinsically linked to apartheid spatial

Figure 1.3 Metrobus in Johannesburg's central business district (photograph by author, January 2020). Metrobus, which is owned and operated by the City of Johannesburg, carries 90,000 passengers daily, making it the second largest municipal bus service in South Africa.

planning – subsidized services were introduced to lessen the financial burden for users who incurred high transportation costs as a result of being forced to travel long distances from their place of residence to their place of employment. The industry peaked around 1982 when it recorded the highest market share; however, by the late 1980s, it went into significant decline, having lost 50 percent of customers to the taxi industry (South Africa Competition Commission 2021). Today, the industry remains in decline even as it collects a guaranteed operational subsidy.

For nearly 20 years, South African transportation politicians and planners have been trying – without success – to transform transportation with BRT. These attempts have yielded eight poorly operating, under-used, or never completed BRT systems: in August 2009, Johannesburg's Rea Vaya became the first BRT system in South Africa (see Figure 1.4); followed in May 2011 by Cape Town's MyCiTi; in 2013, eThekwini announced plans for Go Durban!; in December 2014, Tshwane's A Re Yeng began operating; in August 2015, Go George started in the Western Cape; in June 2018, Ekurhuleni launched Harambee; and in September 2022, Rustenburg started operating Yarona. In 2008, Nelson Mandela Bay attempted to build a BRT but the 60 buses bought for R100 million remain idle due to corruption, poor planning, and politicking (Wood 2022a). None of these systems launched as predicted, with only Johannesburg completing a second phase,

Figure 1.4 Rea Vaya Station in Johannesburg (photograph by author, January 2020). Rea Vaya, which opened in August 2009, is South Africa's first operational BRT system.

and none operate without significant operational subsidies. Hook and Weinstock (2021) explain that just 88 kilometers of BRT have been added as opposed to the 93 kilometers anticipated being built by 2010; and neither Cape Town's MyCiTi nor Johannesburg's Rea Vaya has achieved their proposed ridership – just 70,000 of the projected 100,000 for Cape Town and 56,000 of the projected 165,000 in Johannesburg. As a result, only 30 percent of Johannesburg's Rea Vaya and 40 percent of Cape Town's MyCiTi are covered by fares, resulting in high operational subsidies – R12.84 per passenger (I. Venter 2018). With over R130 billion spent on BRT, this mode of transportation carries less than 1 percent of public transportation passengers in South Africa's six metropolitan areas – 0.95 percent in Johannesburg, 0.24 percent in Tshwane, and 0.13 percent in Ekurhuleni; only Cape Town (1.48 percent) exceeds this threshold (National Treasury 2019).

Despite being the primary focus of transportation planning and expenditure, bus usage continues to decline. Across South African cities, municipal buses carry just 1 percent, private buses 8 percent, the Gautrain 1 percent, Metrorail 21 percent, and taxis 67 percent of public transportation trips (I. Venter 2018). In Gauteng, bus travel declined from 267,066 trips per day in 2014 to just 59,408 in 2019, or just 2 percent of daily trips. At the same time, private car driving as a primary transportation mode has increased from 22 percent in 2014 to 27 percent in 2019. The main reasons for not using buses, according to the 2019/2020 Gauteng Household Travel Survey, were: "no bus available" (45 percent); "not available often enough"

(11 percent); "buses don't go where needed" (9 percent); or "bus stop too far from destination" (5 percent). Figures that might have featured more prominently were: "bus too expensive" (3 percent); "travel time too long or too slow" (0.1 percent); "have to change transport" (0.1 percent); or "prefer taxi" (0.2 percent) (Gauteng Department of Roads and Transport 2020).

Non-motorized forms of transportation – walking and cycling – remain economically and politically marginalized (Mitullah, Vanderschuren, and Khayesi 2017; Wood 2022b) [1; 2; 3] (see Figure 1.5). This is especially important in African cities where as many as 78 percent of people walk every day to access education, healthcare, and jobs. These walkers spend more than one hour each day on foot, navigating speeding cars on unpaved, unplanned sidewalks while breathing polluted air (United Nations Environment Program and United Nations Human Settlements Program 2022). These pedestrians are sometimes referred to as "captive walkers" because they have no transportation alternative [1]. Cycling is also a marginalized mode of transportation in South Africa. Uptake remains low due to economic and social concerns as well as limited infrastructure. Interventions like bike-share have been considered – but not built – as part of policies of transportation transformation and spatial integration (Wood 2022b).

By contrast, the economic, political, and social importance of the taxi industry, South Africa's informal transportation system, cannot be overstated: about

Figure 1.5 Non-motorized transportation in Gauteng (photograph by author, February 2020). In South Africa, as many as 78 percent of daily commutes are made using on non-motorized transportation.

75 percent of total journeys are made by minibus taxis. The growth of the taxi industry mirrors a combination of segregation and anti-apartheid activism as well as Black entrepreneurialism. The system first emerged after the 1930 Motor Carrier Transportation Act, which made it illegal to transport passengers or goods without a permit – a document that was nearly impossible for a Black South African to obtain. The industry grew steadily and by the 1970s the eight-seat Valiants and Chevrolets were common throughout the townships. The 1982 White Paper on National Transport Policy was another milestone – it legalized entry to the market of the now-infamous 16-seat minibus (Wood 2023). As a result, the number of permits issued in 1987 increased from just over 7,000 to nearly 35,000 (Khosa 1991). Today, the South African taxi industry is "a powerful symbol of postapartheid freedom" (Hansen 2012), associated with "cosmopolitanism" (Mutongi 2017) and "indigenous entrepreneurship and creative adaptation" (Agbiboa 2020, 176). It serves 16 million people daily and remains the most popular transportation option in South Africa (Mndebele 2019).

The financing and development of urban transportation in South Africa is also significant as it is reflective of both historical patterns of urban segregation and the failures of postapartheid policies. The earliest forms of urban public transportation – the horse-drawn tram – were only built and operated by private entrepreneurs. In Johannesburg for example, the horse-drawn tram was introduced to transport White miners between their homes in Fordsburg and the mines in Jeppestown (Van Onselen 2001). Apartheid spatial planning furthered these existing practices of segregated transportation. Between 1960 and 1983, when private ownership heavily favored Whites, an estimated 3.5 million non-White South Africans were forced into segregated neighborhoods far from city centers, helping create the sprawling, low-density cities of today (Mead 2016). A separate, segregated bus system was established to transport Black workers between economic and residential locations. To finance two separate bus systems required substantial subsidies. The South African government, like many around the world, therefore developed a complex system of public transportation subsidies to provide services to distant and disenfranchised populaces. And in the postapartheid era, the subsidies, which are said to make public transportation accessible to the poor and the unemployed, linger. The minibus taxi industry, however, remains the only unsubsidized mode of public transportation in South Africa.

Before proceeding, it is important to consider who is attempting to transform transportation in South Africa. Despite efforts to redistribute power in postapartheid South Africa, most transportation planners and politicians remain White men. Of those interviewed for this book, 75 percent were male; 25 percent female; and 67 percent were White, 23 percent were Black, 8 percent Indian, and 2 percent Coloured.[2] These figures are similar to those from my previous book in which 68 percent of interviewees were male and 32 percent were female; 70 percent were White, 14 percent Indian, 12 percent Black, and 4 percent Coloured (Wood 2022a). While there have been some advances in racial and gender equality in postapartheid South Africa – at various moments, transportation in Johannesburg was run by White and Indian women and in Cape Town by Black men and women – further reflection on who is actually planning transportation is necessary.

While this section discussed transportation in South Africa, it is by no means a unique state of affairs. Indeed, infrastructure and systems fail around the world, making what I have just described a highly relatable story about urban challenges and why transportation fails.

Conceptual Contributions

South Africa's experience with transportation fits within a broad set of literature in urban geography and urban studies as well as transportation geography. In this book, these literatures provide the scaffolding for all the chapters and ground it conceptually. This section introduces the two main aims of this book – to contribute to the growing discussion of urban mobilities in Africa and to expand the debates around transportation justice. The chapters intersect with discussions in decolonial geographies (de Leeuw and Hunt 2018; Tuck and Yang 2012), Black geographies (Alderman, Williams, and Bottone 2022; Inwood, Alderman, and Williams 2015), policy mobilities (McCann and Ward 2011; Wood 2022a), walking geographies (Behrens 2005; Zavetoski and Agyeman 2015), animal geographies (Wolch, West, and Gaines 1995; Wolch 2002), art geographies (Bissell and Fuller 2017; Hawkins 2019), and health geographies (Ali and Keil 2007, 2008; Wolf 2016). Consequently, they draw on debates from feminist (McDowell 2016), Chicano (Pulido 2002), Latinx (Faiver-Serna 2019), Indigenous (Radcliffe 2017a, 2018), and queer (Oswin 2020) theorists, whose contributions reveal a more generous, pluralistic understanding of transportation and further the decolonization of transportation knowledge. This expansive approach is central to my argument that transportation needs to draw on a multitude of connected and disconnected, overarching and distinct arguments from architecture, development, planning, and politics in addition to urban geographies. An interdisciplinary perspective therefore reflects the role transportation can play in decentering traditional geographies of knowledge production and decolonizing geography in Africa and beyond.

Urban Mobilities in Africa

First, this book contributes to discussions of urban studies in Africa (Croese and Wood 2024). It builds upon critical engagements with transportation in urban Africa (Pirie 2013; Rizzo 2017) and across the global south (Uteng and Lucas 2017) as well as research into road-building (Goodfellow and Huang 2021), BRT (Jacobsen 2023; Wood 2022a), trains (Van der Westhuizen 2008), and other investments in transportation infrastructure and systems. It constitutes the first multi-modal, multi-sited, interdisciplinary attempt to stretch beyond existing conceptual frameworks to better understand urban transportation in South Africa. In so doing, it contributes to furthering the rigor of African studies and transportation scholarship.

Existing African mobilities scholarship is conceptually critical and empirically diverse. It covers topics from labor exploitation in public transportation (Rizzo 2017) to statecraft in the governance of transportation (Cirolia and Harber 2022); and incorporates a range of intellectual approaches – from cosmopolitanism (Toivanen

2021) to modernity (Green-Simms 2017; Hart 2016), neoliberalism (Rizzo 2017) and policy mobilities (Wood 2022a). It is also geographically expansive – analyzing transportation in Ghana (Hart 2016), Kenya (Mutongi 2017), and Tanzania (Jacobsen 2023; Rizzo 2017) as well as South Africa (Rink 2020; Wood 2022a). Most of the scholarship is explicitly urban, although Porter (2016) studies mobilities in rural Africa; and much of it takes place in Africa's largest cities, except for example Kaur, Forde, and Hayhurst's (2023) study of working-class cyclists in Bloemfontein.

While African urban transportation scholarship analyzes a variety of modes, automobility still features prominently in the literature (Green-Simms 2017; Hart 2016). Public transportation, however, is increasingly the subject of analysis – from bus stations (Stasik and Cissokho 2018) to BRT (Jacobsen 2023; Wood 2022a). Two-wheeled modes like bicycles (Wood 2020a) and motorcycles (Doherty 2022), as well as sustainable modes like mobility-as-a-service (Pollio 2019; Pollio, Cirolia, and Ong'iro Odeo 2023), are also increasingly explored. In many instances, the literature explores beyond the specific details of the mode of transportation. Rizzo (2017), Jacobsen (2023), and Wood (2022a), for example, each use the research on BRT to develop broader arguments related to neoliberalism and policy mobilities. Likewise, Stasik and Cissokho (2018) suggest that both the bus station and the marketplace are "loaded with social, economic, political significance" (2018, viii). For these scholars, transportation highlights the entanglements of people and politics as well as the interconnectedness of African systems.

Informal transportation – the daladalas in Tanzania (Rizzo 2017), danfos in Nigeria (Agbiboa 2022), matatus in Kenya (Mutongi 2017), minibus taxis in South Africa (Nelson 2023), and trotros in Ghana (Hart 2016) – is a common subject through which the exploitation and entrepreneurialism of transportation are often illustrated. These forms of popular transportation generally operate poor-quality vehicles and poor-quality services; they are self-made with low or no governmental involvement or regulation; and they practice high levels of economic competition and violence. The scholarship, however, is attuned to the messiness of informal transportation and the blurry divisions between conceptual and practical understandings of formal and informal. For Rizzo (2017), in particular, informal transportation is neither abhorrent nor blissful but somewhere in between. And for Agbiboa (2022), it is entangled with the "corruption complex." In this book, informal transportation is treated as part of the wider public transportation offering rather than as a distinct mode. Accordingly, the downsides associated with informality are deeply rooted in the failings of formal buses and trains as well as government more generally.

People – bus riders, daladala drivers, motorcycle users, and pedestrians are another subject in African urban transportation literature. Workers, and in particular, drivers, conductors, inspectors, touts, and even policymakers are studied in several of these accounts. Agbiboa's (2022) ethnography of transportation workers in Lagos and Rizzo's (2017) study of transportation workers in Dar es Salaam provide evidence not only of the vitality of urban mobility systems but also of the mistreatments and manipulations by which workers maneuver and are maneuvered by the systems. Meanwhile, other scholars highlight daily commuters and their micropolitics of mobility (Parker and Rubin 2023; Rink 2016). For Parker and Rubin (2023), mobility intersects with gender as parents negotiate their daily commute in Johannesburg (see also Hiralal and Jinnah 2018); while Rink (2016) uses autoethnography to analyze his presence on Cape Town's bus services.

A variety of quantitative and qualitative research methods are used – archives (Morgan 2017), autoethnography (Rink 2016), interviews (Wood 2022a), and mapping (Parker and Rubin 2023), to name just a few. Many of these studies draw on ethnographic evidence from single cases (Agbiboa 2022; Rizzo 2017), but some are also comparative (Sovacool, Daniels, and AbdulRafiu 2022; Wood 2022a). Much of the literature is contemporary, only drawing intermittently on historical happenings (Green-Simms 2017; Hart 2016); however, historical geographies are becoming more common – Morgan and Rink (2023), for example, uncover "the past in the present" in African mobilities.

This book fits snuggly within South African urban mobilities scholarship, which is especially attuned to ongoing transportation challenges. Roger Behrens (2015) and Gordon Pirie (2013, 2014) at the University of Cape Town, Jackie Walters (2013) at the University of Johannesburg, and Christoph Venter (2013) at the University of Pretoria have developed robust studies of transportation practices, policies, and perspectives. Pirie's (1983, 1986) and Khosa's (1995) historical studies are especially noteworthy, providing robust evidence of how and why South African transportation systems developed (see also Morgan and Rink 2023). South African engagements with informal transportation reform (Schalekamp and Behrens 2009; Schalekamp, Behrens, and Wilkinson 2010) and the development of non-motorized transportation options, especially cycling (Jennings 2015a; Kaur, Forde, and Hayhurst 2023; Morgan 2017), also go beyond the usual approaches to transportation.

Transportation Justice

Second, this book aims to weave discussions from transportation and mobilities with social and spatial justice. Its approach builds on Sheller's (2018a) critical analysis of the movement as well as Black geographies (Alderman, Williams, and Bottone 2022; Inwood, Alderman, and Williams 2015) and decolonial geographies (Schwanen 2018a; Wood, Kębłowski, and Tuvikene 2020), and their criticisms of race-led-transportation and colonial-led-transportation. This book furthers these discussions, adding empirical evidence from South Africa in conjunction with conceptual reflections drawn from across geography and urban studies.

This analysis of transportation justice is grounded within Sen's (2009) "capability approach," which draws attention to the unevenness through which different bodies move across space (Pereira, Schwanen, and Banister 2017). "Accessibility, mobility and transport are not ends in and of themselves," Verlinghieri and Schwanen (2020, 2) reason, "but means to ends that are achieved through the activities undertaken across space and time that movement enables." Indeed, the capability approach considers not only movement but also the ways in which potential movement is initiated or restricted based on a range of factors including resources – for example, a car or a bike. This connects transportation justice with broader concerns for decision-making, democracy, inclusiveness, and participation. It also links transportation with the urban agenda and Lefebvre's (1970) "right to the city" as well as Soja's (1989) work on "spatial justice."

The more recent pivot by Sheller toward "mobility justice" (2018a) – alongside her longstanding commitment to mobilities as well as concepts of immobilities and moorings (Hannam, Sheller, and Urry 2006) – calls for action at the micro, meso, and macro scales (see also Sheller 2018b). Verlinghieri and Schwanen (2020) note an important evolution from transportation equity to transportation justice to mobility justice – the latter being constantly on-the-move, shifting and flowing, and the former two being sedentary and fixed. Importantly, mobility justice highlights the ways in which movements and justice are co-produced, historical and relational – and Verlinghieri and Schwanen (2020) add geographical.

There is debate within the literature, however, on how to approach mobility justice. Some scholars draw heavily on Rawls and utilitarianism reasoning that transportation has a social benefit for all users (and would-be users) and therefore investments should increase access for the maximum number (Van Wee and Geurs 2011). In accordance with this approach to equity, transportation planning should benefit the entire system "while ensuring an acceptable level of accessibility of the least well-off group(s) in that territory" (Verlinghieri and Schwanen 2020, 2). For others, the morality of justice is paramount (Lucas, Van Wee, and Maat 2016; Pereira, Schwanen, and Banister 2017). This approach argues that there is an opportunity as well as a need to develop policy alternatives "that maximize the minimum level of primary goods to the people in the worst-off position" (Pereira, Schwanen, and Banister 2017, 175) so that "everybody should be well-off up to a certain minimum threshold" (Lucas, Van Wee, and Maat 2016, 477). For transportation planning, this approach targets the neediest oftentimes at the expense of the whole. Both follow a mobility justice approach and enable city leaders to develop transportation solutions that support the communities utilizing them.

Transportation justice is at the forefront of South African urban scholarship. Debates, however, continue to rage over who should benefit from government interventions and how to allocate limited resources. Before 1994, transportation planning was led almost entirely by White men with limited experiences in public transportation and even less input from users [53; 54]. Since 1994, transportation justice has become a defining feature of transportation planning – in the form of affordable fares, accessible public transportation, shorter travel times, safer modes though with a residual misunderstanding of user experiences or needs (Jennings 2015b). Despite decades of investment in infrastructure and planning, transportation in South Africa is failing. City leaders erroneously emphasize formal and motorized transportation as well as expensive and elite investments rather than everyday usability. To advance transportation justice in South Africa, this book further adds analyses from decolonial geographies, Black geographies, policy mobilities, walking geographies, animal geographies, art geographies, and health geographies. These additional theoretical lenses highlight the politicization of mobilities and its role in conflict and contestations. And they also recognize the breadth of regional and global, human and non-human, visual and visceral characteristics of transportation while still remaining grounded and local. They not only contribute to theoretical understandings of transportation but also provide examples of how city leaders and transportation users can develop solutions that benefit the community utilizing them.

The discussion in this book draws from debates on decolonizing knowledge to expand notions of transportation justice. In doing so, it aims to decolonize transportation by decentering the dominance of northern-based research with a focus on motorized transportation and instead concentrating on unseen (would-be) transportation users and egalitarian modalities. To this end, it explores approaches for reinterpreting transportation knowledge from elsewhere, rejecting knowledge from elsewhere, and refining local knowledge. Indeed, transportation scholars have already begun to engage with these strategies – for instance, in a special issue featuring papers attempting to decolonial urban transportation in the global south (Wood, Kębłowski, and Tuvikene 2020) – but much more needs to be said about transportation planning from an African perspective. This approach provides both a tactile and theoretical perspective on African transportation planning that also reaches beyond the continent.

Similar conversations are also happening in Black geographies where scholarship calls to de-racialize and re-racialize transportation (Hinger 2022). This book furthers these debates by highlighting how geography, a discipline historically detached from issues of race, has recently engaged with Blackness and how bringing transportation into this discussion can further the discourse. To better note the intersections between race, space, and transportation, this book analyzes the policy and legal system of transportation, the community action and protest that mobilizes around transportation, and the informal transportation systems that emerge in response to urban disinvestment and marginalization. In so doing, this discussion provides important first steps for spatial justice and inclusivity.

Urban geographers have been especially attuned to these challenges in their considerations of how and why policies move about and how this movement influences the policymaking process. Like with decolonial scholarship, policy mobilities (McCann and Ward 2011; Wood 2022a) is also sensitive to the disproportionate authority of northern power in knowledge production. For this scholarship, however, the role of mobilized technologies and rationalities is used to analyze urban–global political economies. This book contributes to the policy mobilities discussion by emphasizing the ostensibly ordinary practices and procedures through which transportation models are dislocated and relocated, as well as the ways in which transportation justice is used to develop transportation and urban policy.

Transportation geography has also been central to these discussions. Non-motorized transportation and in particular walking geographies are at the forefront of transportation justice (Behrens 2005; Zavetoski and Agyeman 2015). Rather than further valorizing its role in transportation justice, however, this book brings walking into the discussion to demonstrate the multitude of complex and contested understandings of how to actually enact good transportation planning.

This book also reorients research toward the intersections between animal geographies and mobilities (Hovorka, McCubbin, and Van Patter 2021). It reflects on how animals and humans intersect and intertwine, as well as how animals can serve as mobile actors in the zoopolis. It draws on the movements of animals in South African cities as well as the systematic exclusion of animals and their marginalization from modernity, to further these discussions of transportation justice. It first considers animal movements, second assesses animal policies, and third examines the symbolism and metaphors that a combined understanding of animal geographies and mobilities furthers. This approach contributes a layer of complexity to the conversations in both subfields as well as to the interdisciplinarity of geography.

This book also brings art geographies into the discussion to highlight the cultural significance, social practices, and power relations of transportation as well as the economic and political fabric of the surroundings and the people using it. It notes a complex and controversial relationship between art and transportation, and the ways in which the inclusion of art has been both motivated by politics and profitability. In the African context, however, it highlights the essential role of art in transportation as well as how art on transportation, mobile art, and representations of transportation reveal the inherent mobilities of art. In so doing, it recognizes the ways in which imagery, poetry, and sound negotiate the economic, social, and political power of transportation. This analysis further deterritorializes geography as well as presents a call for spatial and social justice.

A reflection on the intersection between health geographies (Ali and Keil 2007, 2008; Wolf 2016) and transportation is also integral to this analysis. This book considers what happens when transportation fails and how these failures are aggravated during crisis. It explores the overlaps between mobilities and COVID-19 by reflecting on restrictions on all forms of movement including restrictions on international and on urban mobilities. By doing so, it furthers analyses of transportation failure and transportation justice. Indeed, these fields are intrinsically interconnected, although they rarely intersect – divided by journal publications and research approaches, and quantitative versus qualitative research methods. There is an inherent interdisciplinarity to transportation studies and this demands we incorporate scholarship from architecture and city planning as well as politics and sociology, and geography. This book attempts to mix it all up and bring it back together.

Methodologies of Failure

The methodologies adopted in this book, much like the aforementioned conceptual scaffolding, are diverse, drawing on a range of analytical approaches through which to theorize transportation. For African cities in particular, this book engages with "new critical pedagogies – pedagogies of writing, talking, seeing, walking, telling, hearing, drawing and making – each of which pairs the subject and object in novel ways to enliven the relationship between them and to better express life in motion" (Mbembe and Nuttall 2004, 352). It builds upon ongoing discussions of art and aesthetics in mobilities and African cities, and the ways through which formal and informal public transportation systems utilize imagery, poetry, and sound to negotiate economic, social, and political power (Agbiboa 2023; Hansen 2006). My expanded and extended research methodology is therefore especially attuned to the variety of voices involved in planning, financing, and using public transportation as well as those overlooked by existing methodologies.

This study is located in South Africa. However, the methodology remains spatially heterogeneous with chapters drawing on experiences from across the global north and global south (Wood, 2020c). Some chapters are based in one city (Chapter 3), while others move between South African cities – namely Cape Town and Johannesburg (Chapters 4 and 6). Likewise, concepts are drawn from beyond South African cities – transit-oriented development (TOD) (Chapters 2 and 4) and informal transportation (Chapters 2, 3, and 6), to name just two – which although discussed within the South African context, expand beyond any particular location. The (non)locatedness of this study illustrates yet one more way in which transportation scholarship needs to develop new approaches and methodologies.

A total of 55 interviews with senior government and nongovernmental officials in South Africa were undertaken to substantiate this book's findings. The research took place over two periods: between October 2019 and February 2020, I conducted 40 interviews; and another 15 interviews were gathered during a research visit between February and June 2012. Data is drawn from discussions with current and former political and technical heads of city (32 percent) and provincial (6 percent) government departments, consultants (28 percent), academics and research institutes (20 percent), transportation operators (10 percent), and donors (4 percent). The meetings took place across South Africa – in Cape Town (30 percent), Ekurhuleni (4 percent), Johannesburg (64 percent), and Tshwane (2 percent). The majority of meetings were held in person at the participant's office or a nearby coffee shop. Eight took place remotely with respondents based in Cape Town, Johannesburg, London, Manila, Nairobi, and Vancouver.

More than 50 documents are also included in the reference list. I reviewed planning and policy documents from Cape Town (e.g., City of Cape Town 2016) and Johannesburg (e.g., City of Johannesburg 2009), as well as the National Department of Transport (e.g., National Department of Transport 1996), the National Planning Commission (e.g., National Planning Commission 2011), and Stats SA (e.g., Stats SA 2021a,b). The archive of the Johannesburg Development Agency (JDA), a parastatal organization responsible for implementing capital investments, provided essential data for Chapter 5 (e.g., JDA 2015a,b, 2018a,b). Newspaper articles from the *Rand Daily Mail*, a South African newspaper published from 1902 until 1985 are analyzed in Chapter 3 (Staff Reporter 1934a,b) and articles from the *Daily Maverick*, an online South African weekly are used in Chapter 8 (e.g., Diemen 2021; Pikoli 2020). South African (e.g., COWI 2012) and international consultant reports (e.g., Institute for Transportation and Development Policy 2021) as well as press releases from the World Health Organization (e.g., WHO 2020a,b) have also been analyzed. I also evaluated advertising notices and brochures, architectural models and blueprints, art and sculptures, and films and websites.

This book employs a mixture of analytical approaches: content analysis (e.g., interviews) is used throughout the chapters, but it provides the basis for the discussions in Chapters 4, 5, and 7; textual analysis (e.g., newspapers, specifically the *Rand Daily Mail* in Chapter 3 and the *Daily Maverick* in Chapters 7 and 8, as well as government frameworks and reports in Chapters 2 and 5) is used to support both historical (Chapters 2 and 3) and contemporary (Chapters 5 and 7) analysis; and visual analysis (e.g., observations) is central across the chapters (but analyses of art, dance, and sculpture are especially important in Chapter 6). These tactics allowed me to analyze and theorize the infrastructures and influences, the actors and associations inherent to movement (and stillness).

Before continuing, it is important to reflect on my own research positionality as an academic. I write from relatively comfortable and privileged position with a permanent contract, and I am mindful of the ways in which this positionality reinforces the unequal dynamics of urban knowledge production. Yet, I also write from a position of having lived and worked in African cities for years, holding positions in academia, civil society, and local government. I am a female academic joining a growing, but still disproportionate group of female faculty in transportation geography. I am also a lifelong public transportation user. Having grown up in a family that did not always have access to a private vehicle, and then opting not to own a car for much of my adult life, I understand intuitively the challenges of getting to school and work by bus and train. I have also been a captive public transportation user in South Africa. From my personal and professional position, I therefore aim not only to promote Indigenous knowledge about African cities but also to acknowledge the urgent need to address the structural barriers for Africa-based and female transportation researchers to both access and publish academic knowledge.

This book extrapolates from my recurring engagements with transportation in South Africa over the past 20 years. My research of transportation in South Africa began in 2008 when I worked in Johannesburg as a Programs Manager for the South Africa Cities Network (SACN). In 2010, I started my PhD research in London and continued to study transportation across South African cities. My first book, *How Cities Learn* (Wood 2022a), analyzes the adoption and implementation of BRT in Cape Town, eThekwini, Johannesburg, Nelson Mandela Bay, Rustenburg, and Tshwane. It focuses on the initial phases of transportation transformation between 2009 and 2014 and maintains an optimistic tone throughout, an approach representative of the enthusiasm and excitement of the moment. This book adopts a less hopeful tone, again representative of the moment, but one that has been tainted by increases in capital and operational costs, protracted and politicized rollout, and unsolvable operational and infrastructural challenges. However, rather than discussing BRT, it adopts a more expansive approach to transportation, bringing the failures hidden in the shadows to the forefront of transportation planning.

This expanded and extended methodology provides an unparalleled opportunity to critically reflect on our research process. Rarely are researchers able to return to their research sites over such a long period; and even more rarely are they able to interview respondents multiple times. For this book, I interviewed the political and technical heads of Johannesburg's Rea Vaya who were responsible for implementation, as well as having multiple meetings with members of the technical team in Cape Town. Meetings took place in different locations – first in interviewees' government offices but later in their home offices – and in each instance the respondents were at different points in their careers – first as government employees and later as private consultants. Under these conditions, respondents were able to reflect on their personal and professional development, how they worked within and beyond transportation, and how transportation has impacted on their careers. This longitudinal approach allowed me to create a narrative arc of transportation transformation, noting the energy and excitement when it was first being implemented, followed by political transition and concern, and soon after plagued by disappointment after disappointment.

Navigating This Book

This book is organized into nine chapters including this introduction. The remaining chapters build an analytic to explain why transportation fails.

Chapter 2, "Decolonizing Transportation," brings decolonial debates into transportation geography. It considers how geography has engaged with decolonial discourses alongside critical reflections from postcolonial and subaltern studies. These discussions are particularly important for geography, as a discipline historically disconnected from decolonization movements. This chapter then refines its focus on decolonizing transportation by outlining three approaches for undoing and de-doing colonialism: reinterpreting knowledge from elsewhere (e.g., TOD), rejecting knowledge from elsewhere (e.g., bike-share schemes), and refining local knowledge (e.g., minibus taxi). Johannesburg constitutes the context for these considerations, where traditionally transportation has been exploited for colonization and control, but in the postapartheid context is associated with social and spatial transformation. These tactics fit under Johannesburg's pro-poor strategy, through which everyday practitioners and policymakers decolonize the city.

Chapter 3, "Black Transportation," adds a focus on Black geographies to this analysis of urban transportation. Because there is no singular, all-encompassing framework for Black geographies, this chapter draws on a long history of Blackness and Whiteness in the discipline of geography and beyond. It contextualizes the contemporary conversation within critical reflections from feminist, Indigenous, and queer researchers as well as decolonial and postcolonial studies. These wider considerations are especially important for geography, as a discipline historically detached from efforts to de-racialize the city. This chapter then refines its focus on Blackness and transportation by reflecting on race and mobility in South African cities. Three approaches for exploring Black transportation geographies are introduced: the White policies and laws that control movement; community action and protest against racist transportation; and planning for the formalization of the minibus taxi industry. The aim of this chapter, however, is not to promote a particular approach for bringing Black geographies into conversation with transportation, but rather to provide a rigorous reflection that is not only analytically productive but practically useful.

Chapter 4, "People-Oriented Transportation," builds from concepts of TOD, which is generally defined as planned high-density development containing a mixture of residential, retail, commercial, and community uses around a transit hub and surrounded by a high-quality urban realm, that prioritizes the pedestrian (and more recently the cyclist) over the automobile. This chapter analyzes the steps taken in Cape Town and Johannesburg to develop TOD schemes. In so doing, it problematizes both the concept of TOD as a universal mechanism in which all cities apply a similar set of guidelines, as well as the specific planning practices in South African cities. Drawing on the policy mobilities literature and specifically the emerging discussions of policy mobilities failure, I advocate for a people-oriented approach to transportation, noting the challenges and delays in implementing TOD in South Africa. It is not so much that TOD has been applied incorrectly as it has been unable to find its feet in the local context. Accordingly, this chapter presents a fine-tuned analysis of people-oriented transport as both a conceptual framework and a process for actually doing transportation planning. Such a critical reading of the intertwined and overlapping practices of policymaking provides insights into the process of urban development and spatial transformation in South Africa and across the global south.

Chapter 5, "Walking as Transportation," reflects on the concept of walkability, which is generally defined by the extent to which the built environment supports and encourages walking by promoting pedestrian comfort, connectivity, and convenience. This chapter scrutinizes the steps taken in Johannesburg to develop walkability policies and projects. It draws on empirical evidence from recently introduced pedestrian policies (e.g., the Corridors of Freedom) and pedestrian projects (e.g., the Grayston Pedestrian Bridge). Rather than focusing on the merits of walkability or the ease of its application, this chapter highlights the misunderstandings of introducing walkability in Johannesburg. The concept is particularly problematic in Johannesburg, where the "unpleasantness of walking" and "walkability-with-people" are insufficiently considered. In so doing, this chapter aims to first understand walking in an African/global south city and then to decolonize transportation planning by bringing the experiences in Johannesburg to the forefront. Such a critical reading of the intertwined and overlapping practices of policymaking provides additional insights into the processes of urban planning and transportation in Johannesburg and beyond.

Chapter 6, "Trans-paw-tation," relocates these critical discussions in animal geographies by drawing on a range of epistemological and methodological approaches. It goes on to propose three approaches to "trans-paw-tation" and the theoretical possibilities of thinking across animal geographies and mobilities: first, by reflecting on the movement of animals and animals' movements in both the historic and contemporary city; second, by examining animal policies and the ways in which animals are included and excluded from urban mobilities; and third, by considering the metaphorical and symbolic associations between animals and mobilities. In so doing, this chapter moves beyond anthropocentric approaches to mobilities, brings animal geographies into conversation with African and urban studies, and offers a methodological contribution toward understanding trans-paw-tation.

Chapter 7, "Transp-art-ation," brings concepts from art geographies into the conversation. A mural on Johannesburg's main bus station illustrating apartheid history, a painted minibus depicting the living conditions of miners moving through

Cape Town, the dancing taxi guard that became a YouTube sensation, are all forms of transp-art-ation. This expansion of art on transportation represents a shift away from the traditional focus on concrete and asphalt and fluidity and flows. For geography, transp-art-ation contributes to critical visual studies and the ways in which maps and movements expose the economic, social, and political characteristics of urban life. These discussions of art and aesthetics invite further reflection in three specific ways: first, transp-art-ation highlights the broader intersections among art, transportation, and urban geographies; second, it exposes the informality inherent to geography; and third, transp-art-ation promotes the voices of the public in the city. Together, these three contributions underscore the material politics of urban infrastructure in Africa and beyond.

Chapter 8, "When Transportation Fails," reflects on the COVID-19 pandemic and the way in which South Africa became the epicenter in December 2021, when a new variant, dubbed Omicron by the World Health Organization, was identified. This chapter contributes to the burgeoning debates on Covid and the city in two ways: first, it locates the discussions in South Africa, one of the hardest hit countries and one of only a few countries to sequence two variants of concern. Second, this chapter spotlights mobilities and immobilities, uncovering how the pandemic impacted movement and in turn how movement was impacted by the pandemic. The analysis draws on a long history of pandemics, in particular Ebola and SARS, two recent outbreaks that serve as precedents for mobility restrictions including quarantines and international travel bans. From there, this chapter develops a framework through which to understand the South African experience: first by restricting all mobilities, second by restricting international mobilities, and third by restricting urban mobilities. In conclusion, this chapter reflects on what happens when transportation fails.

Chapter 9 outlines this book's main theoretical argument, in particular answering the question of why transportation fails in South Africa and beyond.

Notes

1 Appendix A includes a list of interview respondents including their title, organizational affiliation, place, and date of the interview. Appendix A is used throughout the book to link interview material with interviewees through a numerical system that lists the interviews chronologically. A number in brackets (e.g., [15]) refers to a particular interview and the reader should turn to Appendix A for supplemental information on that source.

2 Racial categories are a legacy of apartheid in which all South Africans were defined according to these four classifications – and these terminologies continue today. I have included the racial categories of my interview respondents to provide the reader with some understanding of the extreme disproportionality between those planning for transportation and those using it.

2 Decolonizing Transportation

Introduction

The theme of the 2017 Royal Geographical Society-Institute of British Geographers (RGS-IBG) annual conference was "Decolonising geographical knowledges: opening geography out to the world." This polarizing focus fueled a series of debates – not all flattering (Esson et al. 2017) – about the continued role of colonialism and empire in geography today. The discussions happening in urban geography (Clement 2019; de Leeuw and Hunt 2018; Jazeel 2017; Radcliffe 2017a, 2017b, 2018; Shaw, Herman, and Dobbs 2006) connect to decolonizing discussions in art history (Grant and Price 2020), Indigenous studies (de Leeuw and Hunt 2018; Radcliffe 2017b, 2018; Shaw, Herman, and Dobbs 2006), literature (J. Gill 2018), theater (Balme and Balme 1999), social work (Gray et al. 2016), and sociology (Connell 2018). These discussions spill into our teaching (Daigle and Sundberg 2017; Wood 2020a). In reflecting on the 2017 RGS-IBG focus on decolonizing, Jazeel suggests "the real value of the conference theme is its assertion that the task of decolonising geographical knowledge is now too important to be left to subfields like postcolonial geography alone" (2017, 334). This chapter proposes to add transportation to this burgeoning dialogue.

A decolonial lens is especially fitting for this study of transportation failure because transportation has long been used to colonize and control. Decolonial scholarship contests European settlement of the New World by tracing debates on modernity and colonialism back to the Portuguese and Spanish invasion in the late 15th century (Mignolo 2007a; Quijano 2007). African colonialism likewise dates to the mid-15th century with the Portuguese conquest of West and Central Africa (Daley and Murrey 2022). Mobilities and transportation were integral to colonialism but their achievements were always in question. Commitments to decolonize transportation by Schwanen (2018a, 2018b) and Wood, Kębłowski, and Tuvikene (2020) add to long-standing calls by Mignolo (2002) and Quijano (2007) to rethink the production and distribution of epistemic knowledge as it relates to transportation and especially as it explains transportation failures. This chapter drills deeper into these claims by highlighting how geography has historically and recently engaged with decolonizing discourses, and how by bringing transportation into this discussion, it can further "see from the south" (Watson 2009).

DOI: 10.4324/9780429322280-2

The next section of this chapter discusses how geography, which first formed alongside colonialism, has thereafter maintained an uneasy relationship with decolonial scholarship. It defines the logic of decolonization, explaining specifically how scholars have engaged with the politics of knowledge production. From there, this chapter situates itself in South Africa where colonial and apartheid logics remain deeply entrenched. It also highlights the particular ways in which decolonial scholarship has been shaped by African scholars. Next, this chapter focuses on decolonizing transportation. It highlights the need and opportunity for transportation geography/ers to contribute toward decolonial discourses. Turning these concepts into action, this chapter then advocates for three approaches to decolonizing transportation in South Africa: reinterpreting knowledge from elsewhere through the localization of bus rapid transit (BRT), rejecting knowledge from elsewhere through opposition to a bike-share scheme, and refining local knowledge through the development of the minibus taxi industry, each of which exhibits a method of decolonizing transportation. The conclusion points to ways to take a decolonizing agenda forward.

Decolonizing Geography

Geography has long maintained an uneasy affiliation with colonialism and imperialism. Early geographers were predominately White, male, elite adventure-seekers mapping and surveying supposedly unchartered territories, classifying cultures, and displaying their antiquities. Their exploits expose the Eurocentric, patriarchal, and racist origins of the discipline. Driver (1992) recognizes the strong links between empire-building and geography, suggesting that it was this association that fueled the growth of geography as an academic field in the first place. The near-absence of disciplinary reflection up to that point is "a sign of the stronghold that the colonial frame of mind has upon the subject" (1992, 26). "It is as if the writings of our predecessors were so saturated with colonial and imperial themes that to problematize their role is to challenge the status of the modern discipline," Driver (1992, 26) concludes.

Opposition to colonial rule spurred a wave of decolonization scholarship beyond geography – Fanon (1961) in politics and Said (1978) in linguistics, for example. In this scholarship, decolonization generally refers to the political campaigns and independence movements happening across African states in the mid-20th century, while decolonial theory by contrast does not only refer to a specific political movement or a precisely defined era. Rather it can also be used as a conceptual starting point to more fully consider the long history of capitalism and control. Fanon explains that colonialism "is not simply content to impose its rule upon the present and the future of a dominated country;" it is "not satisfied merely with holding a people in its grip and emptying the native's brain of all form and content." Rather for Fanon, colonialism "turns to the past of the oppressed people, and distorts it, disfigures and destroys it" (1961, 210–211). Decolonization therefore "sets out to change the order of the world" (1961, 36).

Much of the decolonial scholarship engages deeply with the politics of knowledge. European colonization began "a new world order, culminating five hundred years later, in a global power covering the whole planet" (Quijano 2007, 168). For Quijano, economic and political dominance was accompanied by the coloniality of knowledge, which he calls the "coloniality of modernity/rationality." He advocates for an epistemological decolonization that aims "to liberate the production of knowledge, reflection, and communication from the pitfalls of European rationality/modernity" (2007, 177). Mignolo (2007a) goes on to suggest that the decolonization of knowledge requires not only conceding the geopolitical and historical locatedness of the episteme but also affirms the knowledges disallowed by the hegemony of colonialized forms of knowledge. Mignolo also sees the entanglements of modernity and coloniality reasoning that "there is no modernity without coloniality (2007a, 162)." He argues for the "delinking" of "the rhetoric of modernity" (i.e., growth, development, and progress) and "the logic of coloniality" (i.e., authority and oppression) (2007b). Practically speaking, this means contesting the politics of knowledge production as part of a broader epistemic shift away from Europe as the basis for contemporary thinking.

By contrast, geographers of the postwar era typically neglected decolonization (Clayton 2020; Craggs 2014). Studies of postwar geography reveal a complex imbroglio between radicalism and traditionalism, transformation and hesitation (Craggs and Neate, 2019, 2023). When trying to locate decolonization within geography, Clayton (2020) notes that between 1945 and 1980 the term appears only "a handful of times" in geography journals. To further demonstrate geography's dismissal of decolonization, Clayton (2020) brings the example of the Twentieth International Geographical Union, hosted by the RGS in London in 1964. Of the more than 2,000 delegates, only nine discussed decolonization. Clayton (2020) reasons that geographers were distracted by the everyday revolutions at the expense of the politics of decolonization, focusing instead on issues of development and underdevelopment.

There were however some overlaps between geography/ers and decolonization: British geographers working at colonial universities in Kenya, Nigeria, and Uganda helped to shape both the discipline and the process of decolonization (Craggs and Neate 2016, 2019). Clayton (2020) documents the work of the American geographer Edwin Munger who detailed the independence struggles in the Belgian Congo, British Kenya, and French Madagascar. Munger describes how decolonization simultaneously animated political debate while ushering in violent secessionist struggles. These historical studies suggest a more complicated association between geography and decolonization than typically presumed.

French geographers working in Africa before and after independence likewise did not engage directly with decolonization. They remained apolitical and for the most part conservative. They neither explicitly denounced colonialism nor engaged directly with decolonization. Instead, they committed to a "close and cautious study" of the post-independence actors and associations through rigorous fieldwork (de Suremain 2019, 19). For de Suremain (2019, 19), French geographers working in Africa in the 1950s and 1960s demonstrated the "piecemeal nature of disciplinary change in postwar geography, and how it intersected with changing metropolitan ideas about what geographers thought they could bring to understanding a world in transition." While they remained apolitical, as decolonization in Africa continued French geographers became increasingly engaged with decolonization studies.

One of the first academic books to consider decolonial geographies was *Géographie de la décolonisation* (1971), in which Isnard explored the extent to which independence in French West Africa produced "a geography of decolonization" (de Suremain 2019, 9). Decolonizing for Isnard (1971, 117) meant "giving a country full control of its economy, culture, and social organization: in short, a restructuring of the abandoned legacy of colonization." Drawing on the example of Algeria, Isnard argued for a complete "rupture in geographic space" (1971, 50) because colonialism had so thoroughly disrupted pre-colonial society. Decolonization then necessitated "the invention of an entirely new geography that can respond to the needs and requirements of an emerging nation's authenticity" (1971, 55).

Today, geography is firmly committed to decolonization discussions and praxis. In particular, geography and decolonization intersect through one of three approaches: first, through questions of Indigenous rights and anticolonial movements in the Americas (Adorno 1986; Gruzinski 2002), Australia (Smith 1999), and Africa (Kessi, Marks, and Ramugondo 2020). This approach has been discussed in the preceding section but, specifically for geography, the politics of space and its role in controlling Indigenous land use and ownership, spatial organization, and the spatial layout of colonial settings as well as reflections on the state, power, and identity are all considered.

Second, geographers increasingly use decolonization as a metaphor to challenge hegemonic ideas and practices (Tuck and Yang 2012). This is certainly appropriate given that geography is still dominated by "coloniality-induced institutional racism" (Esson 2018, 2). Desai (2017) calculates the Whiteness of UK geography revealing that while 21.3 percent of undergraduate students identify as Black and Minority Ethnic (BME), in geography that figure decreases to 6.3 percent; within postgraduate studies, only 4.4 percent of geography students are BME as compared to 16.4 percent across the university. Referencing this data, Esson (2018, 5) cautiously advocates for decolonial thinking as well as urban social justice suggesting that decoloniality is a "political project" that "cannot easily be grafted onto pre-existing discourses/frameworks, even if they are critical, even if they are anti-racist, even if they are justice frameworks."

The third focus of decolonial scholarship is on delinking capitalist-colonialist conditions. In following the scholarship of Aimé Césaire (1913–2008), author of "Discours sur le colonialisme" (1950) and founder of the Negritude Movement in Francophone literature, this scholarship adopts a vehemently Marxist tone criticizing the bourgeois capitalistic European culture for stripping the Indigenous of their economic and political agency. To varying degrees, this anti-capitalist thinking draws on the decolonial scholarship of Gaston Leduc (1904–1979). In his analysis of the economic balance sheet of colonialism (1969), he offers a simple definition of decolonization which he characterizes as "a loosening, gradual or sudden, partial or total, of the ties of dependency on the political level." To Leduc, a state is decolonized when it has acquired "full political sovereignty both on the internal level and in the international field" (Leduc 1969, 37). He goes on to recognize that even when political ties are legally severed, an economic dependency remains between the colonial power and the colony. For decolonization to take place then, the economic balance sheet must be rebalanced – a process that has yet to be realized. Meanwhile, Tuck and Yang (2012) add an important differentiation between "external colonialism," used to refer to the expropriation of Indigenous land and resources by a colonial power, and "internal colonialism," through which people and their land are managed by modes of control like prisons, ghettos, schools, and policing. Decolonization under this heading "implicates and unsettles everyone" (Tuck and Yang 2012, 7).

In reflecting on decolonizing as the theme of the 2017 RGS-IBG conference, Jazeel (2017, 334) suggests that "geography has a way to go before it can claim to have transcended its imperial histories." The call to decolonize geography to Jazeel is as "dizzying as it is complex" because it "necessitates as many difficult conversations about teaching and curriculum planning as it warrants critical engagements in our discipline's imperial histories and theory-culture" (2017, 334). This chapter joins these efforts to bring decolonization to the forefront of geography scholarship by emphasizing transportation as the central tenant. In developing a focus on transportation, this chapter also reflects on broader decolonizing methodologies in particular asking how to actually do decolonial geographies. The remainder of this chapter explores this concern, paying homage to the preceding scholarship while also remaining attuned to ongoing debates elsewhere.

Decolonizing Geography in South Africa

In South Africa, the discipline of geography has been profoundly shaped by the country's history of colonialism and capitalism (Visser, Donaldson, and Seethal 2016; Wesso 1992) [5]. "Empire education" was used by British colonial forces to "control ideas" and "colonize the minds" (Wesso 1992, 16). This "Anglicization," "Afrikanerisation," and "Americanization" of geography resulted in an unmitigated disregard for African culture (Wesso 1992, 345). Over time, geography has maintained "a thoroughly colonial sense that imported theories and methodologies are inherently and necessarily superior" (Crush 1993, 62). For most of the 20th century, geographers in South Africa have therefore been reluctant to directly engage with controversial concepts [5]. At Stellenbosch University, "the legitimacy of apartheid ideology rarely was challenged a priori by geographers" (van der Merwe, Visser, and Donaldson 2016, 14) and at the University of Johannesburg, there was "hardly any unrest" as "the student body, for the most part, supported the apartheid state" (Kelson and Kotze 2016, 329). Fittingly, apartheid government departments were filled with geography graduates and geographical approaches were used to embed "race in space" (Crush 1993, 62).

Elsewhere in Africa, by contrast, scholars were expressing vociferous criticism of colonial systems of knowledge. Ake (1979), Santos (1979), and Rodney (1972), to name just a few, ground their critiques of (under-)development within the legacy of colonialism. Ake (1979, i), for example, argues that "one of the most subtle and pernicious forms of imperialism…[is] the guise of scientific knowledge" because "Western social science scholarship on developing countries amounts to imperialism" imposing capitalism and other modes of thought that reinforce the hegemony of the West. While Santos (1977, 87) regarded aid as "nothing else than a form, insufficiently concealed, of the conquest of poor countries by capital; it is a vehicle of domination." Santos specifically blames urban planning for perpetuating colonial systems and norms (Ferretti and Viotto Pedrosa 2018). "It is not even necessary to qualify it as capitalist planning for underdeveloped countries have known nothing else," concluded Santos (1977, 86). For these scholars, Africa's underdevelopment is the byproduct of positivist and technical forms of knowledge that imbue the social sciences with imperialist ideologies. The solution to Santos is an "active and engaged geography" that interrogates the intersections between space and society (Ferretti and Viotto Pedrosa 2018). Decolonization is therefore a "defiant scholarship" (Daley and Murrey 2022), through which to radically revise the core character and ideologies of urban development.

In Kenya, Ngũgĩ wa Thiong'o (1981) argued that decolonizing is a continuous process of "seeing ourselves clearly;" and his tool for delinking modernity/coloniality was to decolonize the mind. "Education is a means of knowledge about ourselves…After we have examined ourselves, we radiate outwards and discover peoples and worlds around us," Ngũgĩ argued.

> The aim, in short, should be to reorientate ourselves by placing Kenya, East Africa and then Africa in the centre. All other things are to be considered in their relevance to our situation and their contribution towards understanding ourselves. … In suggesting this we are not rejecting other streams, especially the western stream. We are only clearly mapping out the directions and perspectives the study of culture and literature will inevitably take in an African university.
>
> (1981, 92)

In practice, Ngũgĩ tried to decolonize by deposing the European imported language, English, instead choosing to write in his native tongue, Gĩkũyũ.

In the 1980s, as resistance to apartheid gained traction, and in particular after the Soweto Uprising, these discussions of decolonization began percolating into geography scholarship in South Africa [5]. Geographers began "questioning the use of many existing models, theories and modes of analysis" (Beavon and Rogerson 1981, 175). Those returning from studies abroad instituted a neo-Marxist critique of orthodox development geography and while this introduced a more critical voice to the discipline, it similarly reinforced a neocolonial reliance on imported models and methods (Nicolau and Davis 2002). In response, Crush (1993, 63) advocated for a "peoples geography" that displaced global north approaches, and "no longer looked to the metropoles for validation, seeking instead to make a contribution to the political struggles against apartheid." Importantly for the arguments I subsequently propose, Crush's approach stressed the everyday practices by which geography could decolonize. This early foray into decolonizing involved showcasing the "hidden geographies" of Blacks and led to a strong wave of historical and urban scholarship.

Transportation geographers in particular responded to this call for a peoples geography by highlighting the everyday experiences of racial segregation and resistance on South African buses (Pirie 1983), trams (Pirie 1989), and trains (Pirie 1992). Of particular importance is Pirie's study of the 1957 bus boycott in Johannesburg's Alexandra township. In careful detail, Pirie provides a historical account of the strike and the role everyday people played in resisting apartheid. Khosa (1992) is similarly attuned to a peoples transportation geography analyzing the growth and development of the taxi industry. Rather than focusing on transportation infrastructure or services, this scholarship exposes the way in which transportation serves as a site of struggle against capitalist, colonial, and racist control. Likewise, it highlights how ordinary South Africans turned the bus, tram, and train into a political and social tool for decolonization.

These decolonial discussions spilled out of the university in 2015 with the #RhodesMustFall (Elliot-Cooper 2017) and #FeesMustFall (Habib 2019) campaigns across South African universities. Students organized demonstrations, marches, occupations, and shutdowns demanding free, decolonized education. Imbued with the writings of Ngũgĩ wa Thiong'o and Milton Santos as well as Black consciousness, radical feminism, and pan-Africanism (Booysen 2016), these movements challenged the "Westernization" of the university and its disregard for African identity and culture (For fuller treatment of RMF/FMF see Ndlovu-Gatsheni 2018). What began as opposition to rising tuition fees grew into a broader mobilization against the enduring presence of colonialism, racism, and inequality in wider society: #fallism extended into the South African government with #ZumaMustFall in which thousands called for the resignation of the then-President as well as protests against statues from the colonial era (e.g., King George V at the University of KwaZulu-Natal) and Anglo-Boer War (e.g., the Paul Kruger statue in Church Square, Pretoria). Fallism consequently exposed not only the fissures in northern-controlled knowledge, or that the majority of research and scholarship continues to be rooted in northern contexts and northern universities, but the broader policies of inequality and exclusion that persist in the postapartheid city.

In recent years, South African geography has taken up the mantle of decolonization, and it is therefore an especially fitting location in which to ground an intervention focusing on transportation. Despite some remarkable decolonial discussions happening in South Africa, in many instances our research remains tarnished by colonial methodologies. And so, while the call to decolonize gains traction, additional discussions of approaches and tactics are warranted. In the next section, I will discuss how transportation can provide a lens through which to consider how to undo and redress colonialism in geography.

Bringing Decolonizing Discourses into Transportation

Some of the most daunting challenges facing cities around the world reside in the transportation sector. Across the global south (as well as the global north), issues of transportation access and inequality are compounded by poverty, unplanned urbanization, and the separation of formal economic activity from residential developments (Priya Uteng and Lucas 2018). These challenges are aggravated by the colonial and postcolonial experience. For this reason, a focus on transportation is warranted within the ongoing conversation on decolonization.

The detrimental legacy of colonialism on transportation in Africa cannot be overstated. Transportation infrastructure facilitated the commercial and economic exploitation of Africa's land, labor, and resources (Debrie 2010; Njoh 1997; Taaffe, Morrill, and Gould 1963). It was used to control the economy by redirecting business to the colonial capitals and focusing on interregional, rather than intraregional linkages (Njoh 1997). Demand for peanuts grown in Senegal and bananas from Guinea for instance warranted the expansion of the ports, railways, and roads in, to, and through Dakar and Conakry (Adeoye 2019). Likewise, Taaffe, Morrill, and Gould (1963) model the process of transportation expansion in Ghana and Nigeria, from a scattering of small dispersed ports along the coast to the growth of major rail lines penetrating the interior. They see colonialism as the motivator for transportation expansion, specifically first and foremost the political or military imperative to connect the administrative centers on the coast with the interior. Mineral expansion and agricultural export production are the secondary and tertiary factors motivating the growth of railways. The railway system in Nigeria (as elsewhere) was not only an "instrument of occupation" built to move resources efficiently from the interior to the ports but also "a tool of exploitation" (Adeoye 2019, 20).

These path-dependent trajectories continue to thwart transportation infrastructure and services decades after independence (Schwanen 2018b); and today African cities remain hamstrung by "transport poverty" with as few as 70 vehicles per 1,000 people, far below the global average of 180 (Agbiboa 2020). Drawing empirical evidence from Cameroon, Njoh (1997) outlines the impact of colonial-based planning on ongoing transportation challenges and notes increased costs to both the state and passengers, the gratuitous cost burden on the poor, increased trip duration for passengers, goods and services, and decreased economic productivity. This has been matched by large-scale transportation infrastructure, typically built and financed by China. From 2000 to 2013, China loaned $13.7 billion to Angola, Ethiopia, and Uganda, amongst others, to build the Belt and Road Initiative as well as international rail transportation (e.g., the Addis-Djibouti Railway) and urban transportation (e.g., the Addis Ababa Light Rail) (Goodfellow and Huang 2021). Like BRT, China's role as a major financier and builder of infrastructure in Africa is debatably neocolonial. On the one hand, it shifts power away from European colonial powers and toward the global south, but, on the other hand, it continues centuries of external presence on the continent.

Geographers have been trying to make sense of these transportation developments. Schwanen (2018a, 2018b) for example has been instrumental in connecting the discussions of decolonization with transportation studies. He acknowledges that transportation studies have "devoted little energy to documenting and critically reflecting on its own history, perhaps because it considered solving the problems that haunt real-world transportation systems more important" (Schwanen

2018b, 4). For the most part, transportation studies excuse the role of coloniality in the development of contemporary conditions. Decolonizing transportation then is about "enabling new modes of problematization and making acceptable other kinds of evidence in debates about planning in cities and states around the world" (Schwanen 2018b, 5). To decolonize, Schwanen advocates for a simultaneous "slowing down and disrupting" of existing expert knowledge.

For Wood, Kębłowski, and Tuvikene (2020), "northern thinking continues to underpin transportation geography, limiting the development of the academic field as well as opportunities for locally derived innovation in diverse localities across the global south and north." In the introduction to their Special Issue, they argue that a decolonial approach to transportation considers models generated in the global south, challenges technically oriented transportation studies, and assimilates informality into the transportation sector. They draw on evidence from Beirut (Samaha and Mohtar 2020), Johannesburg (Wood 2020b), Moscow (Trubina 2020), and Namibia (Baker 2020) as well as existing studies in the global south (Kwan and Schwanen 2016; Priya Uteng and Lucas 2018), in particular from Latin America (Lucas et al. 2018), to suggest a shift in the dynamics of knowledge production. More than merely decentering the locatedness of transportation research, decolonizing involves unsettling traditional centers of knowledge production and the powerbrokers that define transportation problems and legitimize particular solutions. Their inclusion of papers by Samaha and Mohtar (2020) and Trubina (2020) demonstrate strong criticisms of northern discourses of efficiency and sustainability, as well as those by Verlinghieri and Middleton (2020) and Schwanen (2020) illustrating ontologies that originate from outside the established centers of transportation expertise, reflect their commitment to decolonizing transportation.

The third avenue proposed by Wood, Kębłowski, and Tuvikene, that of incorporating informal transportation, has made a significant impact on decolonizing transportation scholarship. A decolonial approach blurs the binaries of formal/informal and regulated/unregulated to redirect attention away from the failings of the state and toward the entrepreneurship of people and place. Informal transportation systems provide evidence of the failures of the colonial transportation systems that were fixated on extracting capital and labor, as well as the vitality of the decolonial condition. Indeed upon closer investigation, informal transportation is fast and flexible to the changing form and function of the city, operates despite bureaucratic obstacles, and is self-financing even while operating on the margins (Olvera et al. 2016; Evans, O'Brien, and Ng 2018). It also provides livelihood opportunities for youth (Porter 2016), women (Naybor, Poon, and Casas 2016), and itinerant workers (Esson et al. 2016) in both urban and rural settings. A decolonial approach to informal transportation would see it as "indigenous transport" (Mateo-Babiano 2016), recognizing it as integral to the decolonial condition.

In this chapter, I consider three additional ways of decolonizing transportation knowledge – by reinterpreting knowledge from elsewhere (BRT), rejecting knowledge from elsewhere (bike-share schemes), and refining local knowledge (minibus taxi). Each of these examples draws from across this literature and debates on the politics of knowledge – from the modernity/coloniality approach by Mignolo

(2002) and Quijano (2007), tools for delinking modernity/coloniality by Ngũgĩ wa Thiong'o (1981), and practices for highlighting Indigenous knowledge by Wood, Kębłowski, and Tuvikene (2020) – to advance a decolonial agenda in urban transportation research. Together these three approaches expose the everyday tactics for decolonizing – be it by decoding or discarding international best practices or by elevating local innovations. In the next section, I will delve deeper into these tactics for decolonizing transportation by drawing on three cases in South Africa.

Decolonizing Transportation in South Africa

In the early 20th century, South African cities were reshaped by British and American transportation models. Transportation "colonialism" begins with the introduction of public transportation services in the White colonial city. Horse-drawn trams were introduced across South African cities – in 1880 in Durban (Jackson 2003), 1881 in Port Elizabeth (Harradine 1997; Shields 1979), 1886 in Cape Town (F. Gill 1961), 1891 in Johannesburg (Shorten 1970), and 1897 in Pretoria (Joyce 1981) – to transport White workers, with exclusive and exclusionary services embedded from the beginning (Van Onselen 1979). Electric trams soon followed the horse-drawn tram – 1896 in Cape Town (F. Gill 1961), 1897 in Port Elizabeth (Joyce 1981), 1902 in Durban (Jackson 2003), 1906 in Johannesburg (Sey 2012; Spit and Patton 1976), and 1910 in Pretoria (Joyce 1981). These systems were primarily designed and developed by private entrepreneurs and, once profitable, were acquired by the city to operate. Like their horse-drawn predecessors, electric trams were the model of excellence in Britain and therefore the assumed solution for the overcrowding and congestion in South African cities (Wood 2022a). In South Africa, however, these systems helped support and sustain systems of separateness and segregation instituted by the colonial government (Wood 2019a).

During apartheid, South African cities were further remodeled by policies of strict racial segregation that denied Black residents full access to the city and its economic base. Blacks were forcibly relocated to residential developments on the outskirts of the city and became increasingly dependent on public transportation to access economic opportunities. The long distances between Black and White, and business and residential uses also had the secondary objective of increasing travel times and costs considerably for Blacks (Wood 2022a). At the same time, apartheid-era planners encouraged car-dependent, low-density sprawl in the White areas of the city.

In the postapartheid era, transportation has been and still is central to the transformation of South African cities. Policymakers are incredibly attuned to the social and spatial challenges that linger and, for the past two decades, they have been working – without success – to develop solutions. A pro-poor agenda is at the heart of these efforts. The examples subsequently discussed represent three approaches of this pro-poor agenda through which South African transportation planners are decolonizing the city: by reinterpreting knowledge from elsewhere – and how a South African model of BRT is being decoded and deciphered, rejecting knowledge from elsewhere – and how a bike-share scheme was vetoed, and refining local knowledge – and how a South African version of informal transportation emerged in reaction to apartheid spatial planning. These examples not only illustrate how South African urban practitioners develop solutions but also how everyday discussions are reflective of a broader decolonizing discourse. Importantly however none of these examples can be deemed successes. Instead, they provide insight into how transportation has failed in South African cities and likewise how these failures need to be more thoroughly scrutinized.

Reinterpreting Knowledge from Elsewhere

This first tactic for decolonizing transportation reflects on the politics of knowledge and the modernity/coloniality approach by Mignolo (2002) and Quijano (2007). Decolonial scholars advocate for an epistemological shift away from Europe as the basis for contemporary thinking. This is particularly important in South African cities that have a long history of mimicking British and European architecture. In this first example, rather than duplicating the international version of BRT, a transportation model that combines the speed and efficiency of a rail system with the flexibility of a bus, South African planners reinterpreted knowledge from elsewhere to develop a uniquely South African version.

Over the past two decades, BRT systems have been introduced in Lagos (2008), Johannesburg (2009), Cape Town (2011), Tshwane (2014), Dar es Salaam (2016), Marrakech (2016), and Accra (2016) with dozens more in the planning stage. These projects have been vehemently promoted by Bogota's former Mayor Enrique Penalosa and American-based NGOs as a global south model ideally suited to global south cities (Wood 2022a). In South Africa, BRT is seen as the solution that simultaneously provides transportation users with an affordable, reliable, and safe transportation system, taxi operators with formalized and stable employment, and buses and rail operators with viable routes. Its larger purpose is to address the severe historical spatial divide along racial lines and postapartheid splintered urbanism. The operational systems conjure up images of equality and dignity for all South Africans, moving freely and efficiently through urban space regardless of skin color or income, in a city free from the grip of informality, and instead managed by an efficient and capable municipal government. In postapartheid South Africa, BRT has become the preferred tool through which to bridge the divided city.

While BRT is not explicitly decolonial, it draws on implicit logics. Transportation interventions in South Africa are driven by a pro-poor agenda that favors captive public transportation users by building affordable and reliable options. "Transport can't be about one level of public transport for the rich and another for the poor," explained Rehana Moosajee, Member of the Mayoral Committee for Transport in Johannesburg from 2006 to 2013 and the political head responsible for implementing Rea Vaya Phase 1A [43]. "We have a lot to do to overcome apartheid spatial planning," another South African politician explained. "We still have the poorest people on the outskirts of the city, the furthest from economic opportunities. There are also all these social divides and public transport is key to overcoming those divides" [54]. Rather than focusing on the efficiency or viability of public transport, these interventions are decolonial. For South African planners, a decolonial approach emphasizes a pro-poor agenda as a means through which to integrate the social and spatial divides.

This decolonial approach to transportation however often leads to associations with failure. The high-capacity of articulated and bi-articulated buses is a feature that leads to failure (Figure 2.1). In the internationally circulated model of BRT, the bus is typically designed to carry high volumes of standing passengers. The South African version of BRT however includes a major social component by providing all passengers with a seat. For Rehana Moosajee, the "dignity of commuters" was of the utmost importance [43]. This is particularly important in South African cities where apartheid spatial planning means that commuters have exceptionally long, arduous commutes. "Part of the problem with existing public transport in South Africa," explains Lael Bethlehem, Chief Executive of the Johannesburg Development Agency from 2005 to 2010, "is that people are treated like sardines." On BRT, "everyone has an individual seat and a seatbelt" [47]. A key metric of success for BRT however is based on high volumes of people moving quickly across the city. South African BRT buses are smaller, with more seats and less standing room than their counterparts in Bogotá: Johannesburg's Rea Vaya operates 143 buses – 41

Figure 2.1 Seats on Rea Vaya bus (photograph by author, January 2020). The seating design of South Africa's BRT buses emphasizes commuter comfort and safety over passenger volumes.

articulated and 103 rigid – carrying between 50 and 130 passengers. The larger articulated bus carries 117 passengers – 61 standing and 56 seated; Cape Town's articulated buses carry 53 seated and 77 standing, the smaller 12-meter rigid buses carry 43 seated and 44 standing, and the 9-meter minibuses carry 25 seated and 25 standing; Tshwane's buses are even smaller – the 18-meter articulated buses carry 90 passengers with 44 seats and the 12-meter buses are designed for 60 passengers. The Bogotá buses are designed to hold considerably more passengers (as many as 270) than those in South Africa. As a result, even if the system reached its projected ridership, patronage would still be lower than initially estimated while operational costs would be higher. It was "a policy decision that for us public transport is about restoring dignity and that is quite different from what happened in Bogota," rationalized Bethlehem [47]. This approach to BRT is rare beyond South Africa.

Rejecting Knowledge from Elsewhere

This second example draws on the reflections by Ngũgĩ wa Thiong'o (1981) and his tool for delinking modernity/coloniality by rejecting external knowledge. For Ngũgĩ, decolonizing is a continuous process of "seeing ourselves clearly," or, in this case, a means through which city politicians and planners clearly saw their urban landscape. In this second example, Johannesburg city officials contemplated and ultimately rejected introducing bike-share – a system of short-term bicycle rental popular in hundreds of cities across the globe. Johannesburg planners carefully assessed their needs but ultimately determined that international best practice is not necessarily local best practice.

Cycling is another of South Africa's decolonizing initiatives driven by a pro-poor postapartheid agenda of social and spatial transformation (see Figure 2.2). However, in Johannesburg, cycling accounts for just 0.2 percent, or one in every 500 trips (City of Johannesburg 2013a). If however the length of the trip is less than 10 kilometers then the modal share rises to 7 percent for low- and medium-income communities (Beer and Valjarevic 2015). In other words, the majority of Johannesburg's cyclists (90 percent) are classified as "striders," otherwise known as leisure cyclists, while the remaining minority (10 percent) are regarded as "stranded" in that they cycle for longer than 30 minutes because they cannot afford motorized transportation (City of Johannesburg 2013a). Johannesburg planners have spent more than a decade debating how to promote commuter cycling and cycling infrastructure – Johannesburg's Integrated Transport Plan (City of Johannesburg 2013a), the Growth and Development Strategy (GDS 2040) (City of Johannesburg 2011), and the Framework for Non-Motorized Transport (City of Johannesburg 2009), all aim to promote commuter cycling. Planners contend that cycling is a pro-poor policy because it mitigates the high costs of motorized transport, a major expense for the poor [2]. One of the primary impediments to cycling, however, remains the initial cost of purchasing a bicycle. Johannesburg planners therefore considered bike-share as a means to increase the availability of and access to bicycles (Wood 2020b).

Figure 2.2 Cycling in Johannesburg (photograph by author, November 2019). Cycling policy in South Africa is driven by a pro-poor postapartheid agenda of social and spatial transformation.

In 2012, officials studied the feasibility of introducing bike-share in Johannesburg. They considered interventions at three locations: between the University of Johannesburg and University of Witwatersrand, between Alexandra (a highly deprived residential area) and Sandton (the City's secondary business district), and between Diepsloot (a highly deprived residential area) and Fourways (a middle-class residential area) (de Beer 2013). "The challenge will be to design a pro-poor bike-sharing scheme which emphasizes bicycles as a means to promote integration and inclusiveness," the feasibility study explained, "and addresses existing spatial inequalities" (COWI 2012, 27–28). The report concluded that Johannesburg's dispersed spatial form, high levels of deprivation, and low levels of tourism could not support bike-share. Postapartheid transformation in Johannesburg is decolonial, with policies and programs attuned to the needs of the poorest residents. This pro-poor agenda meant locating the stations in low-income neighborhoods, instead of business and tourist districts as is characteristic in other cities. This would result in a simultaneous decrease in revenue from ridership as well as advertising income, and operations would therefore necessitate financial subsidies (City of Johannesburg 2014b). The decision by Johannesburg's Transport Department not to pursue bike-share can be attributed to a mismatch between the international model of bike-share and the social and spatial conditions of Johannesburg.

Refining Local Knowledge

The third example of decolonizing transportation draws on the research by Wood, Kębłowski, and Tuvikene (2020) to consider the practices through which Indigenous knowledge is refined and processed. This tactic analyzes the growth and development of the minibus taxi industry in South Africa. In particular, I note the importance of locally generated solutions to locally generated problems and the need to shift theories around good transportation practice away from the conventions of the global north.

The informal public transportation system expanded in South Africa during the 1970s in response to the failures of the state-subsidized, municipal-owned buses and state-owned and state-operated rail services within the townships. During apartheid (not unlike today), transportation was highly politicized, with fierce racial segregation enforced on both buses and trains. Subsidiary systems operated between the townships and the city center, but within the townships, services were essentially absent. Individuals responded by providing a pro-poor, demand-response form of transportation in their eight-seater Valiants (McCaul 1990).

The minibus taxi system, as it was later dubbed, was popular and profitable because it was cheaper, faster, and more efficient than government modes (McCaul 1990) (Figure 2.3). The taxi industry thrived as part of the resistance against the

Figure 2.3 Minibus taxi in Johannesburg (photograph by author, January 2020). Minibus taxis are more responsive and resilient to the needs of commuters making them popular across South Africa.

apartheid state, and it became one of the few avenues for capital accumulation for Black owners and drivers. At first, apartheid officials discounted the industry as a significant threat to other modes of transportation; quite soon, however, the minibus taxi industry started posing a serious risk to municipal transportation services. The 1982 White Paper on National Transport Policy permitted the now-infamous 16-seat minibus to enter the market. While the official motivation for deregulation was to strengthen the emerging Black economy, in reality, it also complemented ongoing efforts by the apartheid regime to destabilize South Africa. Deregulation also interestingly mimicked the global trend of market liberalization. As a result, the number of permits issued in 1987 increased from just over 7,000 to nearly 35,000 (Khosa 1991). This influx of operators led to a series of violent conflicts known as "taxi wars," which have marked the industry since (Dugard 2001). These disputes led to a common assumption that taxis were illegal and operated with impunity. McCaul's (1990) groundbreaking account – very much in accordance with today's decolonial approach to transportation – provides an alternative account of the taxi industry, disclosing the challenges taxi operators faced, as they struggled first against repressive apartheid policy and later against violent mafia-like criminality. Dugard (2001) also roots the cause of the taxi violence and the reasons for its continuation in the uncertain socioeconomic conditions of South Africa's democratic transition.

Consequently, informal transportation or paratransit systems – minibuses, motorbikes, shared taxis, and tricycles – have become the norm across African cities. The trotros in Accra, kombis in Cape Town, daladalas in Dar es Salaam, danfos in Lagos, and matatus in Nairobi account for an estimated 80 percent of total motorized trips in Africa (Agbiboa 2020). In South Africa, today more than 16 million people daily rely on a politically powerful, under-regulated system of minibus taxis – typically 16-seat Toyota Quantum buses. Of total commutes, 65 percent of trips are made by minibus taxi with the remaining 20 percent by bus and 15 percent by train (Mndebele 2019). Informal transportation not only provides users with a cheaper, flexible alternative, but it also forms a significant sub-economy and large numbers of employment opportunities. There are 200,000 minibus taxis in South Africa and each taxi earns an estimated R450,000 annually or about R37,000 per month; while the industry overall generates R90 billion annually, spends R39 billion on fuel, and R2 billion on vehicle insurance. The minibus taxi industry employs 300,000 drivers, 100,000 taxi marshals, 100,000 car washers, and 150,000 vendors at taxi ranks (Fobosi 2020). Accordingly, in South Africa, they are "a powerful symbol of postapartheid freedom" (Hansen 2012), and in Kenya, they are associated with "cosmopolitanism" (Mutongi 2017). To Agbiboa (2020, 176), paratransit therefore not only provides "a means of economic survival" but also serves as "a site of indigenous entrepreneurship and creative adaptation."

Together these examples of decolonizing transportation exemplify a pro-poor agenda aimed at social and spatial transformation. Transportation, which once epitomized the struggle against apartheid, is now the mainstay of urban development – as in the case of the BRT – and is often the subject of experimental interventions – as in the case of bike-share. However, the service most characteristic of the South

African city's pro-poor approach is the minibus taxi industry, a decolonial system that stepped into the void left by apartheid's failure to provide adequate services for Black working classes. In the postapartheid era, the taxi industry has thrived because it appreciates the specifics of the South African topological and topographical landscape and provides the best transportation option for the most users. To summarize, my intention here has been to consider how to actually do decolonial transportation geographies and how these efforts can further decolonize the city.

Conclusion

This chapter contributes to efforts to decolonize geography by developing a conceptual framework as well as a process for decolonizing transportation. It starts by noting how geography, a discipline once strongly saturated with colonialism, is now leading decolonial discussions. From there, this chapter locates itself in South Africa where postapartheid planning has relied on transportation as a tool for social and spatial transformation. It then proposes a focus on decolonizing transportation as one method of actually undoing and redoing colonialism. This chapter then outlines three approaches through which transportation planners are decolonizing transportation: first by reinterpreting knowledge from elsewhere (BRT), second by rejecting knowledge from elsewhere (bike-share schemes), and third by refining local knowledge (minibus taxi). These tactics represent the pro-poor practices through which South African planners are decolonizing the city.

In conclusion, it is important to bear in mind that decolonizing is not without limitations. Indeed, there is debate among scholars regarding the extent to which decolonization discussions fit under the broader umbrella goals of human rights, civil rights, and social justice. Tuck and Yang (2012, 3) for instance remind us that "decolonization is not a metaphor," it is "not a swappable term for other things we want to do to improve societies," and "decolonization doesn't have a synonym." While remaining cautious in mainstreaming decolonization, Jazeel (2017) locates discussions at the intersection of Black, feminist, Indigenous, and queer thinking as well as postcolonial and subaltern studies – reasoning that "the decolonisation imperative means confronting marginalisation will no longer be left only to those who occupy marginal positions within the discipline" (2017, 334). This chapter has aimed to respond to that provocation by outlining an approach to decolonizing transportation that draws on the historiography of the literature on decolonization in general and geography in particular before providing some approaches to decolonial transportation studies. The examples from South Africa highlight the efforts by practitioners to plan better cities, emancipate knowledge from northern epistemologies, and the processual nature of these changes. The aim of this chapter is therefore not to promote only these three approaches to decolonizing transportation or to suggest that these approaches are successful but rather to provide a rigorous starting point for this analysis of transportation failure. Chapter 3 draws on these logics to further a focus on Black transportation.

3 Black Transportation

Introduction

Black geographies scholarship engages with a range of critical reflections from feminist (McDowell 2016), Chicano (Pulido 2002), Latinx (Faiver-Serna 2019), Indigenous (Radcliffe 2017a, 2018), and queer (Oswin 2020) as well as decolonial (de Leeuw and Hunt 2018; Tuck and Yang 2012) and postcolonial (Said 1978) studies. Black geographies also draw heavily on the intersections between these approaches (Driskill et al. 2011; Eaves 2020). This chapter proposes to bring this discussion of Black geographies into transportation studies. Moreover, adding Black transportation to this analysis of transportation failure supports commitments to de-racialize and re-racialize transportation (Bullard 2004; Kwan and Schwanen 2016) as well as longstanding calls to de-racialize the city (McKittrick 2006; McKittrick and Woods 2007). This project considers these claims by highlighting how geography, a discipline historically detached from issues of race and the city, has recently engaged with Blackness and how, by bringing transportation into this discussion, it can shed light on both discourses.

For the broader discussion of failure in South African transportation, this chapter aims to bridge the lacuna between Black geographies and African studies. Scholars of Black geographies have begun to challenge this dissonance – for instance, Pierre's (2013) study of Ghana and Matlon's (2014) analysis of racial capitalism in Côte d'Ivoire – but there is both an opportunity and a need to think about Africa from the perspective of Black geographies. For Hawthorne (2019), the associations between Africa and the Black diaspora create an imperative for Black geographies to partner with African studies. Such a bridge draws heavily on African diaspora studies by providing a counter-narrative to explain modern capitalism. In so doing, it reinforces critical reflections in Black geographies and beyond. Second, this chapter is motivated by calls to bring recent debates on the urban into Black geographies. For Hawthorne (2019), "black matters are spatial matters" because displacement, gentrification, and segregation all seek to exclude Black lives from the urban realm. Issues of (im)mobilities and transportation justice, which leave so many urbanites – particularly in Africa – marginalized (Appelhans, Scholz, and Baumgart 2021; Cirolia and Harber 2022), could benefit from the critical reflections of Black geographies.

DOI: 10.4324/9780429322280-3

This chapter therefore aims to integrate the theoretical approaches of Black geographies with African and urban studies.

The next section of this chapter exposes the long history of Whiteness in geography. This discussion reveals the prejudiced starting point from which Black geographies is launched and explains the imperative to deepen our understanding of Blackness and the city. From there, this chapter focuses specifically on the burgeoning discussions in Black geographies, before turning to this chapter's main aim of bringing transportation into these important discussions. To do this, I advocate for the use of three tactics for considering the tensions among race, space, and transportation: first by studying the policy and legal system; second by considering community action and protest; and third by examining government-led planning. This chapter then locates three examples in Johannesburg first by understanding the 'legalization' of segregated buses via the 1934 Spencer Commission; second by examining public response to racial inequalities via the 1957 Alexandra bus boycotts; and third by tracing attempts to formalize the 'informal' minibus taxi industry. In conclusion, it will be clear how we can take a combined transportation and Black geographies agenda forward.

The Whiteness of Geography

The foundations of racism in geography partly emerged from the scholarship of Immanuel Kant, who in 1755 became the first geography lecturer when he was appointed to the University of Konigsberg (Elden 2009; Richards 1974). There he proffered an inexcusable theory that children in Africa were born White but became Black from exposure to the sun. Kobayashi's (2014) biographical study of the writings of past presidents of the Association of American Geographers (AAG) reveals the deep-seated and long history of racism in our discipline. None of the discipline's early scholars challenged predominant thinking on racial hierarchies. Almond Parkins (1879–1940), president in 1930 and editor of the *Annals*, for example, writes in his inaugural address: "At best the negro slave was a stupid, unwilling, indifferent worker, born with a superabundance of hereditary tropical inertia; and there was little incentive in the institution of slavery to develop diligence, enterprise, and a desire for work" (Parkins 1931, p. 28 in Kobayashi 2014, 1105). Regrettably, Parkins was a highly cited scholar whose scholarship was popular until the 1970s.

The exclusion of Black geographies from Anglophone geography is further evidenced by surveying its scant mention in geography journals. Deskins (1969) was perhaps the first to prove the rarity of research on Blacks: in 1969, he counted just 58 papers about Blacks – just 21 in *Annals* – and these were generally written by White scholars. This survey was expanded two decades later by Dwyer (1997) who counted 176 total articles in *Annals* (2.2 percent of the total number of published articles), *Antipode* (3 percent), *Economic Geography* (1.5 percent), *Geographical Review* (1.4 percent), *The Professional Geographer* (1.2 percent), and *Urban Geography* (8.5 percent). Within this total, 36 percent of articles cover social problems (e.g., poverty, unemployment, crime, and housing), 18.5 percent spatial distribution (e.g., migration, spatial demographics, and mobility), 16.3 percent historical and cultural geography (e.g., slavery), and in 14 percent of the articles, Blacks are incidental to the substantive focus of the paper. Dwyer also notes three peak periods of research: 1911–1963 with a focus on historical and cultural geographies (30 percent of publications), 1964–1980 with a focus on social problems (38 percent), and 1981–1995 with a continued focus on social problems (42 percent) but also including research of the spatial development in residential areas (29 percent). In February 2022, I updated these studies by surveying publications in Annals and Urban Geography. Of a total of 2,487 publications in Urban Geography, just 226 papers include the terms "African Americans" (9 percent); 462 discuss "African Americans" or "Blacks" (18 percent) and 897 discuss "African Americans," "Blacks," or "race" (36 percent). Just 93 papers (3.7 percent) have all three terms. Of a total of 4,976 publications in Annals, 140 include the words "African Americans" (2.8 percent), 327 include "African Americans" or "Blacks" (6.5 percent), and 1,037 include "African Americans," "Blacks," or "race" (20 percent). Only 52 results (1 percent) include all three terms.

The Whiteness of geography is also evident in the demographics of faculty and students at universities in the United States and the United Kingdom. Historically, geography did not feature prominently in the curriculum at historically Black universities in the United States; and by the end of the 1960s, there were just two Black geographers with PhDs teaching in graduate programs in US universities: Harold Rose (the first and only Black geographer to lead the AAG) and Donald Deskins, Jr. (who proved the absence of Black geographies research) (Kobayashi 2014). In 1969, only 12 Black students graduated with geography degrees in the whole United States (Kobayashi 2019). This Whiteness continues in United Kingdom universities today. Desai (2017) demonstrates that, while across the university, 21.3 percent of undergraduate students identify as Black and Minority Ethnic (BME), in geography that figure decreases to 6.3 percent; and for postgraduate studies, only 4.4 percent of geography students are BME as compared to 16.4 percent across the university. This leads Esson (2018, 2) to conclude that geography is still dominated by "coloniality-induced institutional racism."

Even among critical Marxist geographers, there has been little engagement with Black geographies. The writings of David Harvey (1989) and Neil Smith (1984), while furthering an anti-capitalist approach, for the most part, overlook the role of race in the production of space. Harvey (1972, 10) wrote that social change "does not entail yet another empirical investigation of the social conditions in the ghettos." Indeed, he regarded this research as "moral masturbation" by the counterrevolutionary "bleeding heart liberal to prove he is contributing to a solution when he in fact is not." The predominance of class within Marxist criticism shifted attention away from issues of race for several decades. This is a pity since racism and White supremacy are inextricably tied up with capitalism. To prove this, Black geographers like Cedric Robinson (1983) draw on the history of Marxism to illustrate how capitalism and racism co-evolved and are thereafter co-dependent. He applies this concept of "racial capitalism" to explain both specific racial systems like apartheid in South Africa as well as slavery and colonialism as a whole. Black geographies continue to demonstrate "the anti-blackness of capital" and the intersections among race, capitalism, and space (Bledsoe and Wright 2019a).

Whereas historically the racism inherent to the discipline has limited empirical and theoretical engagement with Black geographies, in recent years geography is changing (or at least trying to change) – in 2016, the Black Geographies Specialty Group was formed at the AAG (Eaves 2020); in 2017, a Special Issue on Black geographies was published in the *Southeastern Geographer* journal (Bledsoe, Eaves, and Williams 2017); and in 2018, the AAG highlighted Black geographies as one of its featured themes at the annual conference (Bledsoe and Wright 2019b). These attempts to support the burgeoning discourse are important first steps for a more inclusive geography.

Defining Black Geographies

Black geographies incorporates conceptual thinking from across a range of disciplinary backgrounds and geographical contexts (Bledsoe and Wright 2019b). It does not refer to a specific political movement or a precisely defined era; nor does it define a monolithic Black community. Broadly speaking, "Black geographies" (Gilmore 2007; McKittrick 2006) refers to "black agentic practices of analyzing and actively creating space and place" (Bledsoe 2021, 1017). In their seminal text, McKittrick and Woods (2007) identify three themes across Black geographies scholarship: the othering of Black subjects to disguise their everyday challenges; how the Black community responds and negotiates these geographies of exclusion; and their situated knowledges. By contrast, "geographies of blackness" (Madera 2015; M. Wright 2015) are the "geographic studies of black experiences in space" (Bledsoe 2021, 1017). Black geographies draw on geography as the central analytic through which to understand Blackness, while geographies of Blackness consider race as the focal point from which to consider engagements with space. This is an important distinction, not only between those experiencing Blackness and those actively engaging in the creation of space but also as the analytical explanatory from which to understand contemporary circumstances.

Research of Black geographies has been taking place alongside urban thinking for over a century: in 1899, Du Bois conducted a seminal sociological study of a Black community in the United States. In his study, *The Philadelphia Negro* (1899), Du Bois carefully mapped Philadelphia's Seventh Ward, recording every Black house and business. In following the rigorous empiricism of Du Bois, nearly 50 years later, Drake and Cayton (1945) surveyed the Black geographies of Chicago's Bronzeville neighborhood. They note the machinations through which economic and spatial disparities and racial segregation connect: "Negroes and whites in Chicago are consciously competing for space, for jobs, for political power, and for status" (1945, 758). *The Black Metropolis* (1945) remains a foundational text for Black geographies and urban studies.

Black geographies scholarship increased considerably alongside the American civil rights movement in the 1960s and their demands for economic and political equality (Bledsoe 2021). Liebow's (1967) *Tally's Corner* and Spear's (1967) *Black Chicago: The Making of Negro Ghetto* are two of the important publications from this era, both of which focus on the spatiality of Black life. In *Tally's Corner*,

Liebow responds to the culture-of-poverty thesis by detailing the economic subsistence and cultural happenings of the street corner in the underprivileged areas of Washington, D.C. Like Liebow, Spear also details the sociological characteristics of Black urban life, carefully documenting and describing the rise of the South Side ghetto in Chicago between 1890 and 1920. To Spear, the ghetto developed as a consequence of the "impervious wall of hostility and discrimination which has isolated Negroes from the mainstream Chicago life" (1967, 229). These foundational texts implicitly draw beyond Anglophone geography from the anti-/decolonial writings of Frantz Fanon (1961), Aime Césaire (1950), and the Negritude Movement, adopting a vehement Marxist tone that criticizes the bourgeois capitalistic culture for stripping Blacks of their economic and political agency.

For the most part, Black geographies overlook transportation and mobilities, instead examining Black place-making and the everyday lived experiences of racism across American cities such as Birmingham (Wilson 2000), Detroit (Bunge 2011; Deskins 1972), Milwaukee (Rose 1971), and New Orleans (Woods 2017). Scholars draw on the historical and political processes that led to the production and reproduction of the Black ghetto (Rose 1971), the slum (Bunge 2011), and the (neo-)plantation (McKittrick 2011; Woods 1998). They discuss housing (Rose 1971; Wilson 2000) and other forms of urban development (Shabazz 2015; Woods 1998) as well as incarceration (Gilmore 2007). This scholarship draws on a rich array of interdisciplinary methods – from rigorous quantitative analysis drawing on census data, government records, and maps (Rose 1971; Wilson 2000), to historical analysis and the archives (Wilson 2000; Woods 1998) and cultural approaches, grounded in poetry, philosophy, and theater (McKittrick 2006). It also involves meaningful methods of community engagement and "geographical expeditions" (Bunge 2011; Gilmore 2007). The *Southeastern Geographer* journal has been home to many of these outputs (Bledsoe, Eaves, and Williams 2017; Wilson 1989).

This scholarship is firmly embedded in critical race theory, a conceptual framework challenging how the American legal system approaches racial justice. Emerging in the 1980s out of the failures of the civil rights movement and grounded in the writings of Derrick Bell (1978) and Kimberle Crenshaw (1995, 2010), critical race scholarship reexamines the racial foundations of the US legal system. It also draws heavily on intersectionality (Crenshaw 1991), exposing the connections among race, gender, class, and sexuality. This paradigm shift indicates a movement away from studying instances of individual racism and toward calls to understand the structural conditions of racism (Brand and Miller 2020). Black geographers have applied critical race theories to urbanism by highlighting the racialization of space.

Critical analyses by Harold Rose, a behavioral geographer, one of the first Black tenured professors at the University of Wisconsin-Milwaukee, and the first and only Black president of the AAG, provide another fundamental starting point for Black geographies (Kobayashi 2014). Rose was one of the first to associate racial policy with urban studies, using quantitative approaches and community-led fieldwork to understand the social processes that produced and reproduced the Black ghetto (Rose 1971), Black suburbanization (Rose 1976), and high rates of homicide (Rose and McClain 1990). To Rose, these urban outcomes were the result of generations of Black disenfranchisement, segregation, and exploitation as well as a reluctance and refusal among White people to share space with Blacks. As a result, the "lived experience of black populations in the US was an isolated, segregated, marginalized one" (Bledsoe 2021, 1005).

The scholarship of Bobby Wilson (2000), Clyde Woods (1998), and Ruth Wilson Gilmore (2007) provide another foundation for Black geographies. In their studies of Birmingham, Alabama (Wilson 2000), the Mississippi Delta (Woods 1998), and the State of California (Gilmore 2007), they recognize the systemic disenfranchisement of Blacks from urban life. Drawing on the case of Birmingham, Wilson (2000) reconsiders the mechanisms through which segregation was introduced in the antebellum period and continues into the 21st century. Wilson (2002) incorporates Marxist theories of labor exploitation and surplus value by featuring Black working-class experiences and the exploitative urban spaces in which they lived and worked. Like Wilson, Woods also employs a historical approach to highlight the plantation geographies of the Mississippi Delta. From the pre-colonial period through post-Hurricane Katrina, Woods attributes the economic and social displacement of Black communities to the machinations of a powerful White elite (Woods 1998). Similarly, Gilmore (2007) analyzes the expansion of California's prison system since the 1980s. She associates the exponential growth of incarcerations – nearly 500 percent between 1982 and 2000 – and the disproportionate percentage of incarcerated Blacks and Latinos – two-thirds of the state's 160,000 prisoners – with capitalist exploitation, surplus labor, and racism dating back to the Civil War. These three scholars each highlight the racism inherent in the contemporary capitalist system that results in the criminalization, discrimination, and oppression of Black populations in the United States and beyond. Both Wilson and Woods also identify how Black communities acquiesce to the economic, political, and spatial legacies of slavery, sharecropping, and other anti-Black policies. Wilson also critiques Black activists for neglecting to incorporate class-based struggles into the civil rights movement, thereby enabling existing forms of capitalist-led productions to persist.

While Black geographies has to date neglected issues of transportation, much of the scholarship is explicitly spatial. For McKittrick (2006, xii) and Hawthorne (2019), "black matters are spatial matters." "Space is not just a blank canvas upon which racist activity unfolds," writes Hawthorne (2019, 3), but rather "the production of space is tied to the production of difference." It is just about "identifying the 'where' of blackness in positivist terms," cautions McKittrick and Woods (2007, 6), which "can reduce black lives to essential measurable 'facts' rather than presenting communities that have struggled, resisted, and significantly contributed to the production of space." Indeed, as McKittrick (2006) reminds us, space serves as both the object of study as well as the viewpoint from which to understand racial disenfranchisement. This stands in contrast to the way in which many geographers of the past overlooked the people who actually occupied the space. For Black geographers like Lipsitz (2011) and Shabazz (2015), studies of space reveal and amplify racism and racial inequities. In one of the subfield's most well-cited examples of the intersections among race, gender, class, and space, *Demonic Grounds* (2006),

McKittrick interweaves Black and urban geographies. She traces the geographies of Black women from the transatlantic slave trade to the underground economy, drawing on a range of historical and contemporary facts and fiction. In conclusion, McKittrick draws attention to the longstanding associations between Blackness and spatiality by highlighting "the ways in which racial violences (concrete and epistemic actions and structural patterns intended to harm, kill, or coerce a particular grouping of people) shape, but do not wholly define, black worlds" (2006, 947).

Black geographies scholarship is not a new approach, nor in the process of falling off. However, despite so much impressive scholarship, there remains insufficient engagement with mobilities and movement, concepts inherent to the racial order of contemporary capitalism. Even while mobilities scholarship has proliferated in recent years, Black transportation has continued to be overlooked. And while there has been some attention to transportation in Africa (Cooke et al. 2022; Wood 2022a), transportation planning continues to be dominated by White men and their engineering-led, technocratic, race-blind approaches. A Black transportation geographies lens, as subsequently discussed in this chapter, can provide a fuller picture of injustices in the city and contribute empirically and theoretically to Black geographies.

Bringing Black Geographies into Transportation

Although transportation geography has remained a predominately North American and White discipline, mobilities – forced as well as restricted forms of movement – are inherent to Black geographies. The transatlantic slave trade, the restrictions of the American Jim Crow South, and the Great Migration, as well as the 1896 US Supreme Court decision in Plessy v. Ferguson, are just a few examples in which Black movement has been controlled and cajoled by White systems of capitalism. Indeed, "there has never been a historical period in which the movements of African Americans were not the subject of White control" argue Alderman, Williams, and Bottone (2022, 202). The illogicalities between the locatedness and the dislocatedness of Black bodies, mobilities, and capitalism are at the heart of many Black struggles. Accordingly, Hinger (2022, 368) aptly argues against the "binary of Black = immobile and white = mobile" instead recognizing the complexities of racialized regimes of mobilities. In support of this approach, the subsequent discussion aims to further incorporate transportation into conversation with Black geographies, thereby unsettling the reductive presumptions that render Black bodies as temporary and transient within the city.

Black transportation geographies highlight the racialized regimes of mobilities from automobility (Alderman and Inwood 2016; Alderman, Williams, and Bottone 2022) to public transportation (Alderman, Kingsbury, and Dwyer 2013; Parks 2016) to tourism (Alderman 2018). For automobilities, this approach exposes the freedom and flexibility of car travel as well as how racist restrictions limit movement (Alderman and Inwood 2016; Alderman, Williams, and Bottone 2022). It also highlights the possibilities and limitations of resistance (Alderman and Inwood 2016). For public transportation, this understanding uncovers "transportation racism" (Bullard 2004) and "colorblind transit planning" (Brand, Lowe, and Hall 2020) that disproportionately cost Black public transportation users money, choice, time, and comfort to provide a valuable lens through which to think about the intersections between Black geographies and transportation (Alderman, Kingsbury, and Dwyer 2013; Parks 2016).

Many of these discussions implicitly draw on Cresswell's (2010) politics of mobilities and the entanglement of movement, meaning, and power (Alderman and Inwood 2016; Hinger 2022). Others reflect on the intersections between transportation justice and social exclusion (Lucas 2011). For Alderman and Inwood (2016, 602), "a racialized perspective must recognize that people of color not only challenge racial control by seeking to move on their own terms but are engaged in a complex relationship with oppression." They proffer two terms, "black mobility" to reflect on the ways in which movement is controlled by racist regimes and "anti-racism mobility" to highlight the everyday "countermobility work." Alderman and Inwood's (2016) analysis of NASCAR racer Wendell Scott reveals the possibilities for movement to also serve as an antiractist exercise. Similarly, research into Black travel and tourism (Alderman 2018; Dillette and Benjamin 2022) – specifically the usage of the *Green Book* (Alderman, Williams, and Bottone 2022; Dillette and Benjamin 2022) – highlights the "violent biopolitics of mobility" and the "indivisibility of tourism from the everyday bodily practices of moving through and occupying space" (Alderman 2018, 718).

The overlaps between race and transportation are firmly entrenched in conceptual understandings of space, and it is far beyond the scope of this brief interlude to reference all historical and contemporary intersections (Parks 2016; Seiler 2007). For Bullard (2004), transportation racism is the product of Black immobilities and segregation intended to uphold White supremacy; whereas to Parks (2016), racial immobility sits at the intersection of housing, labor, and transportation. The remainder of this chapter furthers a conceptual and methodological framework through which to bring transportation into Black geographies. To do so, I propose three tactics for considering the tensions among race, space, and transportation – first, by analyzing the laws and policies that produce and reproduce transportation racism; second, by analyzing forms of community engagement and protest; and third, by analyzing the ways in which spatial development and transportation intersect to fuel the growth of informal transportation. Together these tactics demonstrate a process that begins with legal systems reinforcing White supremacy and transportation racism, and then leads to community opposition and protests, and, finally, to the development of informal systems that act beyond the structure of the state. The examples I use initially are all from the United States, but as I will demonstrate the framework is applicable to any Black geographies of transportation.

The first approach considers the ways in which the tensions among race, space, and public transportation are embedded in the US legal system. In one of the landmark legal battles of Black geographies, the 1896 US Supreme Court case, Homer Plessy, a Black train passenger was expelled from a "Whites only" railway car. Plessy argued that separate Black and White carriages were unconstitutional. The court ruled in favor of Ferguson's discrimination policy giving sanction to the "separate but equal" doctrine that would shape American race relations for the next half-century. The ruling was reversed in 1956 when the United States District Court for the Middle District of Alabama ruled that bus segregation is unconstitutional under the Equal Protection Clause of the 14th Amendment of the US Constitution. The state of Alabama appealed and the decision was upheld by the US Supreme Court (Plessy v. Ferguson, 1896; Brown v BoE, 1945; Alderman, Kingsbury, and Dwyer 2013). These two cases provide an example of how race has been central to the Black experience of fixity and flow.

The second approach considers the associations between race, transportation, and community action. The 1955 Montgomery bus boycott in Alabama provides an important opportunity to reflect on how race, space, and transportation have intersected through forms of resistance. In December 1955, Rosa Parks famously refused to surrender her seat on the bus to a White rider. Her arrest led to a 13-month boycott of the Montgomery bus system. There are many interesting analyses of Rosa Parks and her refusal (e.g., Parks 2016); however, Bullard (2004) reminds us that the protestors were not only demanding seats on the bus but also the employment of Black bus drivers, better transportation services to Black neighborhoods, and eradication of the practice of having Black riders pay at the front of the bus and then enter through the back. The fact that Black workers typically have longer commutes, exacerbated by their residential displacement from their workplace, is certainly a part of the story of segregated buses. Likewise, it is important to realize that Blacks are often paid less for the same work as Whites, thereby making the bus, which is usually the cheapest form of transportation, the only option. This protest was instrumental in the rulings of Brown v. Board of Education as well as Browder v. Gayle. Also interestingly, one of the boycott's main organizers was Thelma Glass, a geography professor at Alabama State College, who also served as secretary of the Women's Political Council, the organization that initiated the boycott (Grady 2012). Glass represents one of the first explicit intersections between Black geographies and transportation studies, but she is surely not the last.

The third tactic is to examine the intersection of race, transportation, and informality. Dollar vans – the Grand Marquis, Crown Vics, and Econoline vans that operate in the Black neighborhoods of New York City – are one such example. This informal system picks up as many as 100,000 riders per day and provides a much-needed service to marginalized Caribbean communities (Correal 2018). For Best (2016, 443), these systems "allow black immigrants to inhabit the space through the everyday practice of getting around it." The dollar van routes originated in response to urban divestment in the 1970s and grew substantially during the 1980 transportation union workers strike when drivers capitalized on the lack of other options (Best 2016). In 1993, the city council passed a law requiring van owners to apply for operating permits. The licensing requirements were stringent and expensive, however, causing many to remain unlicensed and illegal (Correal 2018). In response, a schism between licensed and unlicensed drivers (dubbed "plastics") developed and these categorizations continue to this day. Nowadays, dollar vans continue to operate in the subway deserts that result from the failures of the state to provide adequate services (Best 2016). The dollar vans provide some insight into the public response to legal and legislative failures to provide adequate transportation to Black communities.

While scholarship on transportation and race remains interdisciplinary, encompassing discussions in urban planning and urban design as well as geography, it is important to note that much of the existing research is situated in North America. Even as consideration for Black geographies is increasingly prevalent in the literature on transportation in the global north, it remains to be seen how these types of discussions carry through into the global south. The remainder of this chapter is not only empirically located outside of the global north but also draws on a range of marginalized approaches for furthering Black geographies. Indeed, by considering how South Africans interact with transportation, this chapter also highlights the discourses around race in an African context.

Black Transportation in Johannesburg

Blackness is a slippery and elastic concept, and perhaps more so in Africa where Blackness cannot be understood as "singular, essential, ahistorical, given and fixed but as multiple, conjecture[al] and always in process" (H. K. Wright 2016, 86). For Wright (2016, 85), "African blackness" is especially "complex" and "this ought to be taken up more seriously and rigorously in conceptualizations of blackness." In South Africa, Blackness is even more multifaceted – Phadi (2021), for example, draws on the writing of Du Bois and his American experience of Blackness and double consciousness to proffer the notion of "multiple consciousness" for South Africans. In South Africa, Blackness is inextricably linked to the country's history of colonialism and racial capitalism, and attempts to define and delegate race have fueled discrimination and oppression. Systems of racism were exploited for economic, political, and social control and, despite attempts to alter the status quo, these injustices continue to dominate the landscape (Mangcu 2016). The South African lens therefore provides an especially productive and powerful lens through which to ruminate on Black transportation geographies.

Transportation is thus intensely racialized. In Johannesburg, mobilities intertwine with race, class, gender, and other byproducts of capitalism and colonialism (Wood 2019a, 2022a). Colonial and apartheid-era town planners gave little thought to Black transportation, and for decades planners focused on building asphalt highways to accommodate the White elite and their private cars (Wood 2021). Today, both literal and figurative movement is tightly tied to Blackness – Black economic empowerment, postapartheid urban development, and attempts to reform existing public transportation services are all examples of Black mobilities. This section therefore aims to unravel the history of Black transportation geographies as one attempt to better understand Blackness in South Africa.

Legalizing Black Transportation

In Johannesburg, Black transportation geographies began in 1891 with the introduction of horse-drawn streetcars to transport the White miners between their homes and the mines (Van Onselen 2001). Within the year, Johannesburg's Boer government issued regulations segregating public transportation vehicles and excluding Blacks from public transportation vehicles designated for Whites. Then, in 1897, the new colonial government reversed the legislation, permitting Blacks to use tramcars. In 1906, the rules were again reversed and Blacks were only permitted on trams if an alternative service was not available. The all-White council erroneously presumed that keeping Blacks off public transportation would ensure higher ridership figures. For the next two decades, legislation went back and forth reinforcing and reducing segregation aboard Johannesburg's tramcars (Pirie 1989). Finally, in 1927, the by-laws were finalized and from then on Blacks were only permitted to ride in "native" trams (Spit and Patton 1976).

This led to the first intersection among transportation, space, and laws: the 1934 Spencer Commission of Enquiry, which was tasked with determining the future of public transportation in Johannesburg. The Commission pointed out that there is "no city in the world where so large a part of the working population is so badly served with transport facilities" (Staff Reporter 1934a). The Commission went on to argue that the lack of adequate facilities for the Black population was "an unnecessary and cruel hardship," which forced many Blacks to walk long distances to work (Phillips 1935). Spencer, the head of the Commission, explained "that inasmuch as the council has compelled natives to live in locations and possibly at some cost to the ratepayers, it is logically right that the natives should have adequate transport, even if that transport has to be subsidized" (Staff Reporter 1934b). In response, Schreiner, the Chairman of the Johannesburg Joint Council of Europeans and Africa, the agency responsible for representing Black transportation users to the Commission, argued "that the native should be granted permission to use European trams and buses during the slack periods on routes on which no special native transport is provided" and "that the Johannesburg City Council should in some form or other subsidise native transport" (Staff Reporter 1934b). If the city did not provide adequate transportation to Blacks, Fotheringham, Chairman of the Tramways and Lighting Committee from 1929 to 1933, said, "it would be faced once again with private enterprise catering for natives." This might not affect Johannesburg's current financial viability, as it did not cater to Blacks, but Fotheringham noted that segregation was not an indefinite solution – Fotheringham even warned that "the barrier between white and black in Johannesburg was fast diminishing, as in Capetown[sic]" (Staff Reporter 1934a). The Commission recommended "that bus services should be provided along special routes much traveled by natives to outlying suburbs, that special native trams should be run on existing tram routes, and that a restricted number of natives should be allowed to board European trams on routes where native traffic is not heavy, and would cause no inconvenience to European passengers" (Phillips 1935).

The apartheid government exploited the findings of the Spencer Commission to segregate bus services. In Johannesburg throughout apartheid, the city operated ten parallel routes, leading to significant financial costs. Whereas the Black services were typically full and for the most part financially viable, the White services often carried only half-loads and drained the city coffers. In 1978, for example, the

Black fleet of 78 buses carried 16 million passengers while the 365 White buses carried 30 million – less than double the number of passengers with a fleet five times larger. At least, though in response to practices of unequal work and pay, Black fares were often lower than White fares (Staff Reporter 1979).

Protesting Transportation Racism

This leads to the second approach for bringing transportation into Black geographies: community action and protest. During the 1940s, there were eight bus boycotts over escalating fares in Johannesburg and five in Alexandra – a deprived neighborhood north of Johannesburg's central business district, dubbed the "dark city" for lacking sufficient electricity as well as its status as an African township (Pirie 1983). In the 1950s, not unlike today, Alexandra was overcrowded with informal tin shacks, inadequate services, deteriorating roads, underperforming schools, and rampant crime. It was and is also a place of social action where the community comes first and therefore Alexandra is no stranger to bus boycotts (Catsam 2007). In each of the strikes, PUTCO (Public Utility Transport Corporation), the main bus operator for Blacks in Johannesburg, attempted to raise the fare by one penny. The strikes lasted between a few days and several weeks. In each strike, the protestors won and the fare remained fourpence (Pirie 1983). In 1944, a Commission of Inquiry investigating the operations of Black bus services concluded that Blacks could not afford to pay an increased fare – "they certainly cannot afford to pay for anything more in any direction, except by reducing still further their hunger diet" (Khosa 1990, 252). In other words, without an increase in minimum wage rates, Blacks cannot be expected to pay higher bus fares; and because apartheid legislation relegated Blacks to the outskirts of the city, it stands to reason that the government would in turn subsidize their travel costs.

In 1957, PUTCO again announced a one-penny fare increase. In response, under the leadership of the Alexandra People's Transport Action Committee, between January and June an estimated 70,000 residents walked 20 miles to and from Johannesburg's city center to avoid riding the buses (Staff Reporter 1957c). The boycotts quickly spread to Moroka, Newclare, and Sophiatown, three other neighborhoods in Johannesburg, as well as Ga-Rankuwa, Lady Selborne, and Mamelodi in Pretoria (Staff Reporter 1957a). The protest lasted until the government once again agreed to subsidize the fare (Staff Reporter 1957c). While the boycott originated over pennies, it was soon center-stage politics. It happened in an era when Blacks held no right to vote, no representation in municipal or national bodies, no right to assemble, and no outlet for consultation with any authority. Black movement was restricted in every way by "pass laws", which banned Blacks from the city. Despite all these restrictions, Blacks found a method of being heard that echoed across the region and led to substantive change. Accordingly, the success of the 1957 boycott became a fundamental rallying cry for anti-apartheid activities (Staff Reporter 1957b), and many associate the outcomes of the Alexandra bus boycotts with the rise of the African National Congress, the political party that would lead South Africa's first democratically elected government in 1994.

Planning for Informal Transportation

A third approach for reflecting on the overlaps between transportation and Black geographies is a consideration of informal transportation. Chapter 2 discussed the emergence of the minibus taxi industry that transports 16 million people each day – about 65 percent of all public transportation trips in South Africa (Mndebele 2019). The minibus taxi industry provides users with a cheaper, flexible mode of transportation. It also forms a significant sub-economy and a large number of employment opportunities – annually, the industry overall generates R90 billion and the 200,000 minibus taxis each earn an estimated R450,000 (Fobosi 2020). However, it receives no subsidies and pays no taxes. It is organized by two main taxi associations – the South African National Taxi Council (SANTACO) and the National Taxi Alliance (NTA) with more than 1,200 taxi associations operating more than 200,000 minibus taxis (Human and Morrison 2023). In Johannesburg, there are two main taxi associations – Greater Johannesburg Regional Taxi Council and Top Six Taxi Management, managing 61 and 15 taxi companies, respectively (Dlamini 2020). Oftentimes, these taxi associations work together – but at other times, they work in opposition to each other – alongside government attempts to plan for informal transportation.

In the postapartheid era, the government has made some inroads toward formalization. One such initiative was the Taxi Recapitalization Program, first announced in 1999 as a strategy to transform and regulate the minibus taxi industry by replacing old, tattered fleets of 16-seater taxis with modern, shiny 18- and 35-seater vehicles – to be purchased by taxi owners after receiving R50,000 subsidy for each existing van scrapped (South African Cities Network 2011). Although the Department of Transport targeted the recapitalization of 80 percent of the current taxi fleet, by 2007, only R538 million had been spent with another R7.7 billion remaining unallocated (Schalekamp, Behrens, and Wilkinson 2010). The program is for the most part considered a failure – unlicensed drivers still drive recklessly, selectively picking up passengers on impulse, and rider satisfaction remains low. However, for this discussion of Black transportation, it is important to realize that through the Taxi Recapitalization Program the government recognized the significant role of the taxi industry in the transportation network and the national economy and guaranteed their role in future transportation planning (Wood 2022a).

One way in which South African cities have continued to partner with the minibus taxi industry is through the operations of BRT. To many, BRT is yet another attempt to formalize the minibus taxi industry. The process of engaging operators has, however, been much less successful than initially anticipated. Meetings between municipalities and minibus taxi operators included protracted negotiations and many acrimonious meetings. When Rea Vaya Phase 1A began operating in Johannesburg, for example, on August 31, 2009, no agreement had been reached with the taxi industry. Taxi operators remained concerned about the loss of revenue, as their taxi vehicles had to be withdrawn from operations to become shareholders in the new BRT operating company. Another 14 months passed while affected operators were identified, the bus-operating company was established, the distribution of shares was contracted, the reemployment of displaced taxi drivers was arranged, and compensation was provided to those no longer employed (Wood 2022a). On February 1, 2011, the city reached an agreement with the taxi industry and the Clidet shares passed to the new shareholders of PioTrans. Each operator agreed to remove his vehicle(s) from the road and to invest the R54,000 scrapping allowance in the new company as working capital (McCaul 2011). This process remains precarious: each negotiation process is unique, calling for affected operators to debate their concerns and to overcome the resistance among operators to cooperate as well as for the government to determine market share, manage the financial risk, and delegate supervisory responsibilities equally across the system. As a result, this approach to formalization remains intensely problematic.

In a third attempt to formalize and control the minibus taxi industry, the government used the pandemic to introduce subsidies. To mitigate the economic effects of the COVID-19 pandemic, in January 2022, the national Department of Transport announced a R1.1 billion minibus Taxi Relief Fund (TRF), with each applicant eligible for R7,500 (Venkile 2021). Like the previous two attempts at formalization, this approach is not likely to be especially transformative. To be eligible for the TRF, taxi drivers and owners must be South African citizens or permanent residents, in possession of a valid operating license, and registered with the South African Revenue Service (Venkile 2021). As of March 2023, when the TRF application closed, just 85,364 license holders out of 141,987 applicants had been paid. They received a total of R426 million with more than R700 million still left unspent in the fund (Simelane 2024). The extent to which the TRF will be successful remains highly in doubt.

These examples from Johannesburg are not meant to provide a totalizing experience of Black transportation geographies. Nor is Johannesburg emblematic of all cities in South Africa, the continent, or the global south. It is also important to note that these categories are fluid, with policies, protests, and planning intersecting and overlapping across Black transportation in particular and the city in general. Yet, this discussion provides insight into the complexities of failure. It draws on the burgeoning discussions in Black geographies to contribute to wider efforts to de-racialize the city in South Africa and beyond.

Conclusion

This chapter aims to bring urban transportation into the critical conversations in Black geographies. It first reflects on the Whiteness of geography and the foundations of racism from which the discipline developed. From there, this chapter refines its focus on Black geographies, providing a historical overview from Du Bois to McKittrick and in particular considering how contemporary understandings have been shaped by a long record of debate. This chapter then turns to the approaches for bringing transportation into conversation with Black geographies, first by drawing on a range of literature and second by grounding it in South African studies. These discussions are framed around three tactics for understanding transportation racism – first, by analyzing the laws and policies that produce and reproduce transportation racism; second, by analyzing forms of community engagement and protest; and third, by analyzing how informal transportation solutions emerge in response to the failings of the state. These examples present an approach to Black transportation geographies that can serve as both a conceptual framework as well as an empirical process.

It is important to realize that Black geographies scholarship is not without limitations. First, there is a tendency to associate Blackness with urbanity. Black geographies scholarship however demonstrates not only how the city shapes Blackness but also how Blackness shapes the city. This stands in contrast to Whiteness and ruralness. The same could be said for migrants, informality, and of course transportation, which tend to cast a dark shadow on the city. Indeed, the framework for bringing transportation into Black geographies aims to mitigate any presumption that Black equals urban by drawing on legal and legislative considerations as well the community engagements. The role of both state and non-state actors is fundamental in these endeavors, demonstrating not only that these issues are interdisciplinary and interjurisdictional but that transportation racism negatively impacts all people living within and beyond the city limits.

Second, there is a propensity to rely too heavily on historical studies as evidence that Black bodies belong in the city. This is a double-edged sword: it is a dangerous tendency, not only because of the history of colonialism and slavery but also because it constrains contemporary research. However, it is also a necessary means through which to counteract Whiteness and racism. This approach for bringing transportation into Black geographies therefore takes account of this dilemma by thinking relationally between examples in an attempt to highlight the analytical possibilities for bringing thinking on Black geographies into conversation with transportation.

In conclusion, this chapter has outlined an approach to transportation failure by drawing on Black geographies scholarship before providing some approaches to actually doing Black transportation. In so doing, it has both bridged the lacuna between Black geographies and African studies and between Black geographies and urban studies. The examples from Johannesburg highlight not only the efforts by academics and practitioners to build better cities and de-racialize transportation but also the protracted nature of these conversations. Rather than promoting these particular three approaches to Black transportation geographies, this chapter seeks to establish a reflective interlude that is both analytically productive and practically useful in understanding transportation failure. Chapter 4 contextualizes these reflections within a discussion of people-oriented transportation.

4 People-Oriented Transportation

Introduction

People-oriented transportation (POT) is a South African version of transit-oriented development (TOD) adapted to reinterpret ostensible transportation failures as successes. TOD is defined as "integrated urban places designed to bring people, activities, buildings and public space together, with easy walking and cycling connection between them and near-excellent transit service to the rest of the city" (ITDP 2017, 8). For TOD to be successful, urban planners develop new tactics that encourage density and diversity by designing walkable mixed-use precincts centered on public transportation. Rather than segregating residential, retail, and commercial uses as has been standard in cities around the world, TOD seeks to mix economic, environmental, and social interests with urban forms. And almost every city worldwide has adopted some form of TOD as part of generally accepted practices of good urban planning (Suzuki, Cervero, and Iuchi 2013). This chapter analyzes the steps taken in Cape Town and Johannesburg to develop TOD. It is not a case of successful implementation, however; instead, this chapter considers the ways in which TOD has not gone according to plan and how the concept has been adapted in South African cities into POT. In exploring the failings, this chapter problematizes both the concept of TOD as a universal mechanism as well as the specific planning interventions in South Africa.

This chapter locates the development of POT within the "policy mobilities" (McCann 2011; McCann and Ward 2011) literature. Research of the movement of best practice – like TOD – from locality to locality as well the process of mutation – from TOD to POT – has expanded in geography and urban studies. The scholarship scrutinizes the topographical movement of policy actors and their physical travels (McCann 2011; Rapoport and Hult 2017) as well as topological comparisons made through benchmarking and rankings (Bunnell 2015; Robinson 2015). Research of the circulation of other transportation best practices such as bus rapid transit (BRT) (Wood 2015a) and bike-share (Wood 2020a) are also worth noting. This chapter aims to frame TOD explicitly as a mobile policy moving around the world through a process of learning, dialogue, and politically motivated exchange. Rather than assessing the outcome of a successful instance of policy adoption, this chapter reflects on the failings of policymaking and the

DOI: 10.4324/9780429322280-4

challenges of introducing transportation policies from elsewhere as well as the mechanisms through which these failures were reinterpreted as a South African version of TOD called POT.

The next section of this chapter aims to problematize concepts of TOD. I trace the history of transit-oriented thinking before introducing two major gaps in the literature, research of "bus-oriented development" and "people-focused development." From there I examine the policy mobilities approach, outlining the broader questions of how and why best practice circulates. Bringing these two literatures together, I turn to the empirical discussion of the two case studies, the Klipfontein Corridor in Cape Town and the Corridors of Freedom in Johannesburg. These examples illustrate a South Africanized version of TOD that may provide lessons for cities around the world. In conclusion, I reflect on the main theoretical arguments that serve to problematize concepts of TOD.

Problematizing Transit-Oriented Development

It has been more than 30 years since Peter Calthorpe, the American architect famous for promoting concepts of new urbanism, coined the term "transit-oriented development" (TOD). In his book, *The Next American Metropolis* (1993), Calthorpe defines TOD as "a mixed-use community within an average 2,000-foot walking distance of a transit stop and a core commercial area" (1993, 56). He goes on, "TODs mix residential, retail, office, open space, and public uses in a walkable environment, making it convenient for residents and employees to travel by transit, bicycle, foot or car." The concept had its first trial in 1988 in Calthorpe's study of Portland, Oregon in which he convinced the city to introduce a light rail, instead of the proposed Western Bypass highway, and to cluster housing, offices, and retail around the stations (Gauvin 1994). For Calthorpe, this was not so much a particularly new idea as a call "to reinvent the old streetcar suburb, where you had fabulous downtowns and you had walkable suburbs, and they were linked by transit" (Kunzig 2019). He calls himself "a reviver rather than an originator of ideas" (Newman 1991).

Indeed, transit-oriented theorizations appear throughout urban history and these historical developments continue to shape contemporary framings of TOD (Knowles, Ferbrache, and Nikitas 2020). American streetcar suburbs (e.g., Brookline outside Boston) and British new transit towns (e.g., Bedford Park outside London), where land development was facilitated by the electric trams, serve as antecedents (Knowles, Ferbrache, and Nikitas 2020). The most famous predecessor is the garden city, an urban planning model popularized by Letchworth (completed in 1919) and Welwyn Garden (completed in 1920), two new towns built on the northern outskirts of London in the early part of the 20th century (For a fuller treatment of Letchworth, see Lewis 2015). Interestingly, the garden city concept also spread around the British Empire, for example, the development of Pinelands, a suburb of Cape Town, South Africa, which opened in 1919 (Wood 2019a). In each of these developments, the train station and its associated commercial precinct served as the focal point (Hardy 1992).

Contemporary understandings of TOD as a financial tool can be traced to American transportation agencies where declining ridership in the 1970s gave rise to the need to develop real estate around the stations to cover operating costs (Belzer et al. 2004). These practices of "transit-related development" (Dittmar, Belzer, and Autler 2004, 6) turned into what Cervero (1993) dubs "transit-supportive development" as transportation agencies turned into developers throughout the 1980s. Land-based financing expanded and transportation agencies soon extended into the neighborhoods beyond the stations enabling a form of "development-oriented transit" (Dittmar, Belzer, and Autler 2004).

A clearer definition of the terminology is needed before proceeding. At the broadest level, the label TOD can be used as a planning instrument to promote principles of inclusivity, livability, and sustainability; whereas in a more prescriptive sense, it refers to stand-alone, mixed-use buildings adjacent to transit hubs. Both planning and academic literature use this nomenclature to promote sustainable development. They emphasize the symbiotic relationship between land use and transportation as well as the central role of decision-makers working in the economic, environmental, and social sectors across governmental and nongovernmental agencies. Cervero and Kockelman (1997) highlight the importance of the three "Ds" – density, (i.e., concentrations of uses around transportation stations in for e.g., Tumlin et al. 2003), diversity (i.e., mixture of residential and retail uses in for e.g., Schuetz 2015), and design (i.e., pedestrian-friendly and cycling facilities in for e.g., Ewing and Bartholomew 2013) – and later they add an additional three "Ds" – destination (i.e., regional accessibility in for e.g., Renne, Hamidi, and Ewing 2016), distance (i.e., access to the train station and walkability in for e.g., Akbari et al. 2018), and demand-management (i.e., reduction of parking spaces in for e.g., Ewing and Bartholomew 2013) – for increasing transportation ridership while decreasing vehicle miles traveled. In spite of the many studies of the six "Ds," Jamme and others (2019, 409) note the absence of the "D" in TOD which, while it typically signifies land development, has become universalized.

In Africa, principles of TOD and transportation corridors have been applied in Ethiopia (Institute for Transportation and Development Policy 2021), Ghana (Agyemang et al. 2020), Kenya (Onyango and Owino 2021), and Tanzania (Ka'bange, Mfinanga, and Hema 2014) as well as in South Africa – in Cape Town (Cooke, Behrens, and Zuidgeest 2018; Wilkinson 2006; Wilkinson and Marks 2007), eThekwini (Markewicz, Linders, and Moodley 2011), and Johannesburg (Harrison et al. 2019; Wood 2021). These analyses identify and assess the long-term and short-term economic, environmental, and social outcomes for TOD. eThekwini's "The Northern Urban Development Corridor," for instance, aims to achieve higher densities by directing development into concentrated nodes along major transportation routes (Markewicz, Linders, and Moodley 2011). In many instances, however, the application of TOD has been partial, incomplete, or retroactive without achieving the promised outcomes. For Cooke, Behrens, and Zuidgeest (2018), the relationship between development and transportation is not entirely understood. In Cape Town, they explain, residential densities and public transportation usage are already high but much of this happens in informal developments and by informal transportation. While in Johannesburg, Bickford and Behrens (2015) attribute the limited achievements of TOD to a range of conflicting and confusing objectives by both the state and local stakeholders.

Two areas relevant to the forthcoming discussion of people-oriented development are "bus-oriented development" and "people-focused development." First, the synergies between transportation, land, and density are generally attributed to development along railways and around rail stations rather than busways and bus stations. One exception however is Wohlwill (1996) who observes new subdivisions comprised of single- and multi-family houses as well as shopping centers, entertainment, and medical facilities built alongside busways in Pittsburgh, Pennsylvania. In light of global proliferation of BRT in the past two decades (for details see Wood 2015a), it is surprising that so little attention has been paid to "bus-oriented development." Likewise, the literature overlooks the association between TOD and non-motorized mobilities such as cycling and walking. Both would entail a thorough reconsideration of bus and bicycle services as high-value, fixed infrastructure worthy of adjacent investment and land redevelopment. Research of bus-oriented development might therefore shed light on the way in which buses, as well as bicycles and walking, can stimulate TOD.

Second, for the most part, the association between TOD and "people-focused development" is sidelined. A people-focused development approach puts the needs of the community first by involving urbanites in the planning process. While community-building was central to Calthorpe's (1993) definition, around the world TOD has fueled gentrification and displacement by increasing land values (Chatman et al. 2019). Jamme and others (2019, 409) note a shift in TOD research from a "utopian definition centered on communities to an operational and site-specific definition applying to land use only." And over the past 25 years, only 12 percent of references recognize an association between community development and TOD (Jamme et al. 2019, 413). Likewise, Chatman and others (2019) conclude that while there is an established correlation between income and automobile usage, the association between income, automobile usage, and proximity to public transportation remains nebulous. Accordingly, research into people-focused development would consider who benefits from TOD and how planners can use TOD to encourage more equitable development practices and universal access to public transportation.

While the scholarship on TOD is interdisciplinary, encompassing discussions in urban planning, transportation, and urban design as well as geography, much of the literature on TOD draws heavily from the North American experience. And even while practitioners and planners typically attribute their learning to the example of TOD in Curitiba, Brazil (see the Klipfontein case study later in this chapter), much of the academic discourse remains focused on the global north (Knowles, Ferbrache, and Nikitas 2020). A third contribution, then, is to theorize not only the way in which the TOD concept has been applied outside of the global north but also how cities of the global south serve as prototypes of TOD. This more global orientation allows for opportunities to critique the concept and consider its universal applicability.

Bringing Policy Mobilities into Transportation

Research of "policy mobilities" (McCann 2011; McCann and Ward 2011), the process by which policy moves across jurisdictional boundaries, has ballooned in recent years. Studies of the global proliferation of best practice (Moore 2013), public policy (Peck and Theodore 2015), transportation (Wood 2022a), and urban management (Baker et al. 2016) have been particularly fundamental in recent theorizations. The scholarship identifies the mobility of policy knowledge and models (Montero 2018) via policy actors (Rapoport and Hult 2017; Wood 2014) and their policy organizations (Theodore and Peck 2011; Wood 2019b) as well as across sites of learning (McCann and Ward 2012; Wood 2015a) interpreting best practice as moving both topographically via policy actors and their physical travels (Rapoport and Hult 2017; Wood 2014) and topologically through benchmarking and rankings (Wood 2020b). This literature can explain the extensive efforts to promote TOD in South African cities, and, in turn, the empirical examples can expose new ways of understanding policy mobilities.

The policy mobilities rubric is a particularly useful lens for analyses of South African cities. This is not the first instance in which South African cities have been used as evidence for policy mobilities research, however – Didier, Peyroux, and Morange (2012) analyze the mobilization of business improvement districts and Wood (2022a) traces the introduction of BRT in South Africa to South American cities. These studies highlight the connections between cities in South Africa and both the global north and global south and counteract misunderstandings of South African cities as unique (Harrison 2015; Wood 2015a). This chapter adds to these studies by highlighting the circumstances through which TOD knowledge flows across South African cities as well as the unintended consequences of that knowledge flow.

The policy mobilities literature engages with failure through the concept of "best practice," a "powerful heuristic tool for the dissemination of innovation and knowledge" (Moore 2013, 2371). For Peck and Theodore (2010, 171), best-practice policy models "travel with the license of pragmatic credibility, and models that emanate from the 'right' places invoke positive associations of (preferred forms of) best practice." Moving away from the assumption that best practice provides a template for universal reproduction, the policy mobilities literature reveals the way in which the policy introduced differs from the one at its site of origin (Montero 2016). Imported policies are often reshaped to suit the needs of the importing locality. González (2011), for instance, considers how urban regeneration models mutate to fit the needs of a wide array of adopting localities; and Robinson (2011) concludes that policy models are always subject to negotiation at the local level. Yet, still relatively little is known about how these mutations take place. This chapter highlights the features that attracted South African policymakers to TOD and in so doing reflects on the theoretical notion of best practice and the ways it mutates into local practice.

The scholarship is especially attuned to concerns of policy failure. Harris and Moore (2013) ask what it means when best practice fails to transfer, in reference to research by Crot (2010) on the challenges of transferring models of participatory budgeting from Barcelona and Porto Alegre. Harris and Moore (2013) suggest that in this instance policy actors may have had incomplete knowledge, which limited their ability to act. In my research on BRT adoption, I reason that what appears as a case of failure may instead be a case of slow policy transfer; when it returns by drumming up stronger support for a second attempt to apply the knowledge (Wood 2015b). In another study of the circulation of the garden city model, I use policy mobilities literature to illustrate the "four-fold instance of failure" (Wood 2019a): the first failure is noted as an instance of incomplete transfer by which international transfer agents spread their interpretation of planning concepts rather than a prescriptive planning model; the second serves as the precursor to bad planning – in this instance apartheid's spatial segregation; the third shows how failure is experienced by the architects and planners introducing the garden city model whose professional lives were stunted afterward; and the fourth failure is noted as a missed opportunity to develop local urban solutions and systems that might have served the city better in the future. Failure in South Africa, as in many places, may be the result of technical challenges in applying the TOD model or it may result from political contestation over responsibility for implementation. Or perhaps TOD has not failed entirely because its precise progression remains uncertain throughout the learning and adoption process. This chapter engages with these concepts of failure to understand how and why TOD was introduced in Cape Town and Johannesburg as well as its failings.

It is also important to consider the role of the actors shaping policy mobilities failure. Prince (2016, 424) notably makes an appeal to "avoid overly fetishizing the actual movement of policy" and instead "pay more careful attention to the specific circumstances in which policy is adopted." In my research on the circulation of BRT, I have argued for the central role played by local policy actors in adopting mobilized policies (Wood 2014). Such contentions are particularly useful in understanding the way in which the locality captures and grounds policies-in-motion. This chapter contributes to policy mobilities discussions by deliberating on the practices through which best practice from elsewhere is approved, embraced, and/or disregarded by local actors and on their interactions with global advocates and inter-referencing across space and time. To better address these concerns, the next section explores the processes by which concepts of TOD were introduced in Cape Town and Johannesburg.

People-Oriented Transportation in Cape Town and Johannesburg

There is a strong association between land development and transportation in South Africa. The mining outposts and port villages of South Africa, like their European counterparts, separated the worker from their place of work and the factories from the inner city, all of which necessitated the development of a viable means of public transportation. The story of TOD in South Africa therefore begins with the inauguration of horse-drawn streetcars, in Cape Town in 1863 (Gill 1961) and in Johannesburg in 1889 (Van Onselen 2001). Policy mobilities logic is implicit within these developments. Architects frequently pursued antidotes to local urban challenges in British town planning models, which diffused through the global hegemony of Western imperialism and were implemented with relative ease (Lemon 1991). In the early 20th century, for example, the British garden city model, which called for efficient, decentralized, residential areas to replace the dirty and crowded cities of industrialized England, was replicated as Pinelands in Cape Town (Wood 2019a).

It is not surprising then that South African politicians and planners root their support for TOD within concepts of density, growth management, and inclusiveness, which they learned from the Brazillian city of Curitiba in the years following democracy. Capetonians and Joburgers traveled to Curitiba to see TOD projects firsthand and they returned eager to install these best practices back home (e.g., City of Cape Town 2003). Others learned about TOD while studying or working in Europe and the United States [41; 51; 53; 55]. One engineer in Cape Town [55] explains,

> Transit-oriented development goes back here for me to the 1980s when I was at Berkeley doing my graduate work…That was new then and in 2003, it was new here. It became the flavor of the month but it was donkey's years old.

South Africans remained committed to the principles of good public transportation and called upon these earlier encounters in the planning of both the Klipfontein Corridor in Cape Town and the Corridors of Freedom in Johannesburg.

Klipfontein Corridor in Cape Town

In 2002, Klipfontein Road, a double-lane road running from the southern suburbs of Mowbray and Athlone to the townships of Gugulethu, Nyanga, and Khayelitsha, was slated to be Cape Town and South Africa's first TOD corridor (see Figure 4.1). Klipfontein was identified as Phase 1 of Cape Town's Mobility Strategy which laid out a framework for public transportation corridors that integrated mixed land uses and higher densities. Recalling the best practices of Curitiba, the plan was to use BRT to transform the 20-kilometer stretch of desolate road into an economic heartland. The Klipfontein BRT would run in the median lane, and the unimproved land along the Corridor would be redeveloped as convenience stores, supermarkets, coffee shops, crèches, hardware stores, and newsagents to serve the tens of thousands of BRT users (City of Cape Town & Provincial Government of Western Cape 2005). The then-Provincial Minister of Transport in the Western Cape, Tasneem Essop, imagined Klipfontein Corridor becoming a destination:

> We want to inject life, inject hope into the Klipfontein Corridor … We want to turn it into a social and cultural hub again. … Imagine them [township residents] taking an evening stroll to a street café on a paved street mall that is well-lit and secure.
>
> (Williams 2003)

Figure 4.1 Klipfontein Road in Gugulethu, Cape Town (photograph by author, February 2020). Klipfontein Road, a double-lane road running from Mowbray to Khayelitsha, was scheduled to be Cape Town and South Africa's first TOD corridor.

The Klipfontein Corridor project failed for various technical and political reasons. Technically, the project was a challenge because the two-lane road through Mowbray was particularly narrow and there was limited space to reallocate to BRT; politically, the community protested, claiming that the City wanted to tear down historic buildings. While the plans could have been altered to accommodate the narrow streetscape, overcoming the engineering challenges would have required significant political and community support [48]. And as challenges mounted, the partnership between the Province and the City fizzled. In 2003, the City was permitted to plan for public transportation initiatives while the Province was responsible for funding the infrastructure. When the City failed to match the financial investment of the Province, the relationship fell apart [52]. And in 2004 Essop, who had been driving the project, was reassigned to another portfolio and her TOD project fell by the wayside [45].

But the project did not entirely fail; rather it morphed into the plans for the MyCiTi BRT in Cape Town. Even after the Klipfontein project was canceled, the local actors continued to promote TOD. Lynne Pretorius from the City and her counterpart in the Province, Zaida Tofie, two local actors who had been involved with Klipfontein, founded a consulting firm specializing in public transportation and together they continued to work on transportation projects. They were appointed in 2008 to do the operational planning for Cape Town's MyCiTi Phase 1.

And the plans for TOD along Klipfontein remained on the books with the City's Department of Transport. In 2016, Klipfontein Corridor resurfaced in the City's TOD Strategic Framework as one of ten corridors that utilize public transportation infrastructure to redress the spatial legacy of apartheid (City of Cape Town 2016). Maddie Mazaza, Director of Transport in Cape Town, explains that the MyCiTi Phase 1 shifted from Klipfontein to the R27 and Klipfontein became Phase 2 [54]. And at present, the City is developing the business and operational plans for a Klipfontein transit corridor. The earlier iteration of the Klipfontein Corridor project was therefore not a complete failure. On the contrary, it was instrumental in creating both the physical and political space for ongoing TOD projects.

Corridors of Freedom in Johannesburg

In May 2013, Johannesburg's then-Mayor, Parks Tau, launched an ambitious program of TOD, dubbed the "Corridors of Freedom" (Tau 2013). The plan promised to reshape the future of the city by linking transportation interchanges with high-density, mixed-use residential, retail, and business developments as well as educational and leisure facilities (City of Johannesburg 2014a). Three corridors were selected: the central business district to Soweto (a former Black township to the southeast) along Empire and Perth Roads which follows Rea Vaya BRT Phase 1B (City of Johannesburg 2012a); Louis Botha Avenue between the central business district and Alexandra along Rea Vaya Phase 1C (City of Johannesburg 2012b) (see Figure 4.2); and Turffontein node, a working-class district south of the central business district (City of Johannesburg 2012c). Proposed interventions included upgrading parks, clinics, libraries, and sports facilities as well as creating pedestrian crossings and cycleways. And in reference to the best practices in Curitiba, a tenfold increase in population density was also recommended, from an average in Johannesburg of 4,000–7,000 people per kilometer-squared to about 40,000 people per kilometer-squared and a morphology of 10–12-story buildings immediately adjacent to the stations and 4–6-story buildings in the surrounding areas [14] (see Figure 4.2). The city set aside R1.5 billion over a period of ten years (or about 15 percent of the city's R10 billion capital budget) [6] as well as an additional R2.6 billion for capital investment in transportation systems such as Rea Vaya BRT (City of Johannesburg 2014a). In response, private developers were expected to bankroll infill rather than peripheral development (Todes and Robinson 2020). And, the goal was to transform entrenched settlement patterns, which have shunted the majority of residents to the city's outskirts, away from economic opportunities.

The project did not unfold as expected: in August 2016, Johannesburg elected a new mayor to lead a coalition government of the Democratic Alliance and the Economic Freedom Front. The new mayor, business tycoon Herman Mashaba, argued that these interventions focused too heavily on the mostly White northern suburbs at the expense of the inner city and the predominantly Black southern townships. One planning official recalls that Mashaba "didn't want to ride on the Corridors of Freedom bus, which was such a strong brand for Parks Tau" and "he was more orientated towards immediate or daily service delivery with less of a focus on the

Figure 4.2 New developments along Louis Botha Avenue in Johannesburg (photograph by author, January 2020). Louis Botha Avenue, running between Johannesburg's central business district and Alexandra, and home to Rea Vaya Phase 1C, is one of three Corridors of Freedom.

future growth of the city and how you can take the pain now, but reap the benefits later" [14]. The project, which had been both under-funded and under-resourced throughout, hit a wall (Robinson et al. 2020). And despite incentives for developers, only a few came on board while the majority remained focused on building along the periphery (Todes and Robinson 2020). Robinson and others (2020) moreover suggest that bus-based transportation is a minor component of travel in Johannesburg – with developers preferring to invest around the region's Gautrain rail system – and yet the Rea Vaya BRT was meant to serve as the backbone of this mega project. And while the project trudged along under the generic nomenclature of "Transit Corridors," to a large extent the momentum was lost, and by 2018 the project was defunded.

While the Corridors of Freedom has been renamed and its significance downgraded, the longstanding commitment to align development and public transportation continues. In an effort to better understand the successes and failings of the project, Robinson and others (2020) consider the response of private developers to public investment, and Harrison and others (2019) assess units of inclusionary housing. And while it is too soon to fully comprehend the long-term results of the Corridors of Freedom – with the plan intended to unfold incrementally over several decades – four planning shifts are worth noting: first, by 2017 there were at least 60 applications from private developers eager to consolidate and convert residential and commercial parcels into higher densities [14]. These developments include around 5,000 units of affordable housing. This leads to the second result: in 2019, Johannesburg became the first city in South Africa to implement an inclusionary housing policy that requires developers to allocate 30 percent of new housing as affordable (Robinson et al. 2020). New student accommodation on the Empire-Perth corridor is a third associated result (Harrison et al. 2019). The final metric is one that fits the categorization of development-led transportation: the number and frequency of Rea Vaya buses in operation have increased. The F2 bus route now runs 18 buses per hour, up from 14; bus numbers on the C4 route have doubled to 16; and the C2 has increased its complement from 18 buses to 22 per hour [17]. More than a story of success and/or failure of TOD in South Africa, the Corridors of Freedom provides insight into a South African version of TOD, a topic to which this chapter now turns.

A South African Approach to Transportation Planning

The South African experience with TOD offers three pathways for geography and urban studies. First, the South African circumstances demonstrate that POT is not a prescriptive mechanism for development. Indeed, there is no single approach that covers all possibilities or provides for all opportunities. Across sectors and spheres of government, South Africans use the phrasing of TOD in reference to "inclusivity" [46], "accessibility" [48], or "social cohesion" [45]; or it can address improvements in public space or "liveable spaces" [14]. In South Africa, concepts of TOD have morphed into concepts of POT which combine economic and community needs. According to one Johannesburg-based planner, TOD "is about how you plan the city and so in the planning you may not even call it TOD, but in fact it influences the way you think" [46]. As one of the key planners of Johannesburg's Corridors of Freedom explained, "I don't really like the term TOD, because of the amount of misconceptions that it creates as well. At one stage the city manager kept on saying, 'We should build a new highway because it's TOD.'" For Johannesburg as elsewhere, he concludes, TOD is both a "brand" and a "style of planning" and for the Corridors of Freedom, it was "the tipping point and the spine around which we could restructure and rebuild a new kind of city" [14]. However, the application of these concepts in practice was more aligned with the South African spotlight on inclusivity and transformation.

Second, POT does not only materialize along fixed and/or costly transportation lines such as rail or even BRT; informal transportation systems can also provide opportunities for development along corridors of movement. One of the limitations of TOD in South Africa has been the tendency for city leaders to concentrate investment around fixed routes without taking account of the possibilities for building TOD along informal transportation nodes. The minibus taxis, though not strictly formalized, have particular routes and predictable stops and city leaders could find ways to stimulate capital within the existing transportation systems rather than fixating only on investment along the BRT route. Likewise, non-motorized forms of transportation could be considered as studies suggest over 29 percent of all peak period trips are made on foot (Gauteng Department of Roads and Transport 2020). One of the limitations for both the Klipfontein Corridor in Cape Town and the Corridors of Freedom in Johannesburg is that the government tended to concentrate investment alongside high-capacity systems like rail, presuming that fixed infrastructure was more permanent and long-lasting and therefore necessary to court private development. As rail is under severe pressure in South African cities, it is increasingly apparent that other modes of transportation be considered for strategic planning.

Third, the South African experience reveals that POT can happen through both macro- and micro-transactions. This stands in stark contrast to the internationalized version of TOD which calls for large-scale development and high-end economic investment, for the most part ignoring the importance of the frequent micro-transactions of the everyday. Economic transactions happen wherever passengers board/alight; and even micro-transactions such as buying repackaged crisps or single cigarettes from a snack stand generate economic externalities. In Johannesburg, for example, a TOD corridor could have been implemented in Soweto between Meadowlands and Dobsonville, which is a taxi-only route with a noticeable degree of informal transactions and trade [17] (see Figure 4.3). The

Figure 4.3 No trading zone at Metrobus bus stop in Johannesburg (photograph by author, March 2014). A people-oriented approach to transportation provides opportunities for business and trade at bus stops.

snack stand that folds up at the end of the day, the barber shop that consists of a chair and a sign, and the cook selling lunch to resting taxi drivers are strategically situated within transportation nodes. Indeed, these small-scale entrepreneurs and micro-transactions are integral to the broader transportation economy and should be considered within TOD.

Conclusion

This chapter proposes to problematize the concept of TOD by applying policy mobilities logic to transportation planning in South Africa. The empirical examples show how a seemingly straightforward application of TOD in South Africa can twist and turn with far-reaching and unexpected consequences. TOD did not entirely fail in either Cape Town or Johannesburg; rather it reformed into other projects – such as when Klipfontein Corridor became MyCiTi Phase 2. For policy mobilities scholarship, this underscores how policymaking failures are generative, yielding unanticipated opportunities for better urban planning. Policy mobilities is more than the outcome of the topological and topographical movement of ideas but also a process of adaption and mutation facilitated by local (South African) policy actors and their interactions with elsewhere. This suggests a further need to challenge related concepts of best practice by dispelling the impression of an easy application of international ideas. No program or policy can ever be a *best* practice, only perhaps a *better* practice, and only in certain times and places. And in highlighting the politics of these decisions, this chapter also exposes the failures inherent to practices of policy mobilities.

For transportation, this chapter further complicates the notion of TOD by challenging calculative studies of transportation geography. Rather than seeing TOD as a mixture of residential, retail, and commercial development situated along a fixed public transportation system, as might be typical in transportation studies, this chapter takes a critical perspective advocating for TOD as a tool, a starting point rather than an end, for building inclusivity, livability, and sustainability. The South African experiments provide three openings through which to rethink TOD scholarship: first, TOD is a fluid, malleable notion, and one that takes account of the existing circumstances – in fact, TOD is deemed a failure when it does not take account of the setting into which it arrives. Second, the South African understanding helps us realize the possibilities of "bus-oriented development," "taxi-oriented development," and other associations between informal modalities and land value. Third, the empirical discussion directs academics and practitioners toward a greater appreciation for the micro-transactions that fuel the transportation economy yet are often sidelined in traditional TOD literature. Indeed, it is the imbrication of these three provocations that invites further research in African cities and beyond.

For the broader discussion of transportation failure, this chapter confirms the importance of locally made decisions. Rather than focusing only on the successful circulation of ideas from here to there, this chapter considers what actually happens when a policy arrives. The successful implementation of mobile policy often requires local policymakers to unfold and refold it to fit within the prevailing local context. The examples from South African cities therefore direct transportation planners – and academics as well – to consider applying not only the expensive fixed transportation systems implemented elsewhere but also the informal systems, small transactions, and moments of encounter happening locally. Indeed, these experiences in South Africa help us to generate a more complex understanding of transportation failures both within the continent and around the world. Chapter 5 considers these findings within an analysis of walking-as-transportation.

5 Walking-As-Transportation

Introduction

For Walter Benjamin, the flâneur was an affluent resident of 19th-century Paris who aimlessly strolled the city for entertainment (Lauster 2007). Such a romantic image stands in sharp contrast to the contemporary pedestrian scurrying across six lanes of fast-moving traffic to get to their low-paying job, or the seller trudging along the unpaved, unimproved verge taking their wares to the market. The purpose of this chapter is to engage with the concept of walkability – generally defined by the degree to which the built environment supports and encourages walking by providing pedestrian comfort, connectivity, and convenience (Southworth 2005) – and assign it a priority status alongside other modes of private and public transportation. In so doing, this chapter applies a transportation justice lens to walking by elevating its status as a primary transportation mode. It draws on empirical evidence from the pedestrian policies and projects recently introduced in South Africa. In addition to demonstrating the merits of walkability and its application in South Africa, this chapter also highlights the challenges, limitations, and failures of actually doing walking-as-transportation.

This analysis of walking-as-transportation is increasingly pertinent in African cities, where pedestrian rates are higher than in their American or European counterparts. On average, 50 percent of trips in African cities are made on foot – with as many as 81 percent in Dakar, 70 percent in Kinshasa and Addis Ababa (Pendakur 2005), and as few as 53 percent in Niamey (Olvera et al. 2010) and 42 percent in Ouagadougou (Pendakur 2005). Walking, for many Africans, is their only means of transportation – especially for the young and the elderly members of poorer households (Behrens 2005). Many of these trips last nearly an hour as people negotiate dangers from the air, from moving cars, and from unimproved pathways (United Nations Environment Program and United Nations Human Settlements Program 2022). Rapid urbanization, urban sprawl, and the role of government as both a transportation policymaker and policy regulator, further exacerbate already limited mobility choices (Pirie 2014).

Walking is especially noteworthy in Johannesburg: of the 3.5 million daily trips, 800,000 trips are made by walking ("all the way") (City of Johannesburg 2013a). These pedestrians are "captive walkers" because they have no alternative

DOI: 10.4324/9780429322280-5

mode of motorized transportation (Montgomery and Roberts 2008). In Johannesburg, 66 percent of households do not own a car (City of Johannesburg 2009); and 60 percent of households spend more than 10 percent of their income on motorized forms of public transportation, up from 55 percent in 2014 (Gauteng Department of Roads and Transport 2020). These captive walkers are also "time poor" with the average pedestrian exerting significantly more time and energy on mobility (Montgomery and Roberts 2008). Over the past five years, walking time to the first public transportation stop has increased from 9 to 14 minutes, while access to the final destination has increased from 8 to 14 minutes (Gauteng Department of Roads and Transport 2020). Moreover, pedestrians account for 40 percent of all road fatalities (Road Traffic Management Corporation 2018).

This chapter analyzes the practice by which city officials in Johannesburg introduce walking-as-transportation policies and planning. It contributes to the theoretically rich discussions of walkability (Zavetoski and Agyeman 2015) as well as empirical surveys of walking in African cities (Behrens 2005). The next section reconceptualizes walkability across the urban studies scholarship. From there, this chapter locates walkability in African cities and considers how it has been developed and studied in various contexts. The following section then turns to a discussion of walking-as-transportation in Johannesburg focusing specifically on the way in which walkability is unequal. From there, this chapter focuses on the failings of walkability in Johannesburg, citing evidence from the Grayston Pedestrian Bridge, the Westbury Pedestrian Bridge, and the Milpark Pedestrian Bridge. These examples provide valuable insight into South African misunderstandings of walkability. In conclusion, I reflect on the main theoretical arguments that serve to problematize the concept of walkability.

Reconceptualizing Walkability

Walkability is a central characteristic of urban space with an equivocal definition (Dovey and Pafka 2020; Forsyth 2015). Scholars across a range of disciplines have attempted to define the concept – in health studies, walkability is attributed to the physical attributes of the local environment (Leslie et al. 2005; Rigolon, Toker, and Gasparian 2018); in transportation studies, walkability is correlated with household travel behavior (Manaugh and El-Geneidy 2011); and in urban studies, walkability sits at the intersection between economic and social development and urban design (Dovey and Pafka 2020). This chapter relies on a multidisciplinary approach, following Southworth's (2005) definition of walkability as

> the extent to which the built environment supports and encourages walking by providing for pedestrian comfort and safety, connecting people with varied destinations within a reasonable amount of time and effort, and offering visual interest in journeys throughout the network.
>
> (Southworth 2005, 248)

The physical characteristics of the urban environment at both the microscale (e.g., sidewalks and crosswalks) and the macroscale (e.g., residential density, variety of business) influence the behaviors of pedestrians (Rigolon, Toker, and Gasparian 2018). The quality of pedestrian facilities, roadway conditions, land use patterns, and security provisions are important components of walkability (Ewing and Cervero 2010; Litman 2003). Walkable areas are "close" (i.e., a short distance between origin and destination), "barrier-free" (i.e., accessible for all bodies), "safe" (i.e., void of crime and cars), "full of pedestrian infrastructure and destinations" (i.e., sidewalks, raised pedestrian crossings, street furniture, trees), and oftentimes "upscale, leafy, or cosmopolitan" (i.e., "pleasant for middle-class professionals who have other choices for getting around") (Forsyth and Southworth 2008, 2). Also importantly, some of the physical factors of walkability include the extent to which the walkable paths are connected (i.e., in a comprehensive network of pathways), convenient (i.e., direct routes without lengthy diversions), comfortable (i.e., no steep gradients, good weather, good lighting, feeling safe), convivial (i.e., the presence of landscaping and street furnishings), and conspicuous (i.e., legible routes with clear signage, integration with public transportation) (Pharoah and Buchan 1997).

Walkability is often seen as a measurable indicator, with various associations attempting to codify a set of metrics for assessing walkability: as early as 1997, the US Department of Transportation developed a walkability checklist to allow parents and children to characterize the walkability of their neighborhood (US Department of Transportation 1997). In 2006, Krambeck (2006) proposed a qualitative analysis of walking conditions called the Global Walkability Index that included a public survey. In 2008, the World Bank proposed another five dimensions – accessibility/mobility (i.e., the distance and duration of the average walking trip), safety (i.e., the rate of pedestrian fatalities per population), security (i.e., the rate of pedestrian-targeted crime), legality (i.e., pedestrian laws and rights), and public expenditure (i.e., percent of total urban infrastructure and maintenance funds spent on pedestrian facilities) – to measure the pedestrian environment in cities of the global south (Montgomery and Roberts 2008). In 2018, the Institute for Transportation and Development Policy, a New York-based sustainable transportation advocacy organization, introduced *Pedestrians First: A Tool for Walkable Cities* (2018). This set of 11 indicators considers walkways, crosswalks, visual frontage, physical frontage, shade, and connectivity as essential characteristics of walkability. These attempts to assess the walkability of an urban area and its associated infrastructure are flawed, however, because the metrics are generally difficult and cumbersome to measure. Likewise, while it may be straightforward to measure the walkability of a neighborhood, it is more complicated on a city scale. Moreover, these indices fail to adequately consider why people walk, focusing too heavily on the characteristics of the pedestrian environment.

In addition to these features, walkability is also used as a planning instrument to promote principles of inclusivity (e.g., the association between walkability and equity), livability (e.g., the association between walkability and the local environment), and sustainability (e.g., the association between walkability and land use) (Litman 2003). For Litman (2003), walkability has far-reaching economic, social, and environmental benefits, by increasing employment, generating tax revenue, raising property values, decreasing transportation costs, and reducing the amount of land used for roads and parking. Planning for walkability therefore requires an alignment between "residents' desires and expectations for types of destinations, their willingness to walk a given distance and the quality of the required path" (Manaugh and El-Geneidy 2011, 309). Given this extensive list of benefits, furthering walkability is often considered to be just good transportation planning.

Most of the scholarship on walkability responds to the proliferation of car dependency and suburban growth; and it generally advocates for a reduction in car usage as well as increased infrastructure and amenities to support walking, cycling, and public transportation (Khayesi, Monheim, and Nebe 2010). Some of this scholarship is inspired by Jane Jacobs (1961), the New York-based urban activist, whose observations of urbanism remain the cornerstone for walkability (Dovey and Pafka 2020), while, in other instances, scholars take an engineering slant. Buchanan (1963, 40), for instance, recognizes the inherent conflict between traffic and the environment as well as the "accessibility problem"; and Whol and Martin (1967) expose the economic and social inadequacies of the then-current transportation research. Appleyard (1981) added research on "livable streets," a term he introduced to address both the macro and micro characteristics of the roadway. In his study of street morphology in San Francisco, Appleyard demonstrated a link between traffic volume and sociality concluding that residents of streets with low traffic volumes had three times more friends than those living on streets with high volumes. This understanding of "the street as a sanctuary" (Alca˘ntara De Vasconcellos 2004, 8), what Appleyard called the "ecology of the street," emphasized the relationship between mobility and livability or the driver and the resident.

Walkability is generally understood as a feature of urban development with studies employing a range of terminology: "universal design," a term used for a dropped curb on an accessible sidewalk (Goldsmith 1963); "road diets," when a four-lane road is converted to three (Labuschagne and Ribbens 2014); and "complete streets," a term coined in 2003 by Barbara McCann and the American advocacy organization America Bikes to incorporate crosswalks, sidewalks, bicycle lanes, and road medians into mainstream transportation infrastructure. As opposed to "routine accommodation," the lexicon previously used for bicycle-based transportation planning (Zavetoski and Agyeman 2015), a complete street includes cyclists, pedestrians, motorists, and "transit riders of all ages and abilities" in transportation planning, ensuring that they are "able to safely move along and across a complete street" (McCann and Rynne 2017, 3). A complete streets policy then considers the whole right-of-way (e.g., sidewalks, crossings, cycle lanes, and road medians) in the design, planning, operations, and maintenance of the road. The concept was picked up in the United States by the American Planning Association and American Association of Retired Persons, as well as by international advocates including the Bicycle Empowerment Network based in Cape Town and Embarq in Mexico City (Prytherch 2018). Such terminology, according to Zavetoski and Agyeman (2015), denotes a form of "street-level spatial justice," a "democratization of the street" achieved by decentering drivers and redistributing the right of way to the public; and Litman (2015, 14) suggests that complete streets policies "help achieve equity objectives by giving non-drivers a fair share of road space." The concept of complete streets is not without criticism, however: Zavetoski and Agyeman (2015), for instance, propose that such language dehumanizes road users and presumes an unjust rationality to modal choice. Moreover, it presumes streets exist in the abstract by isolating the designated area from the other parts of the city.

While scholarship on walkability remains interdisciplinary, encompassing discussions in urban planning, transportation, and urban design as well as geography, it is important to note that much of the existing literature and best practice is situated in North America even as the concept has increasing international relevance (Guerra, Cervero, and Al 2017; McCann and Rynne 2017). The abovementioned definitions of walkability are generally drawn from examples in the global north, and these classifications and conceptualizations are not always relevant in the global south. In these northern contexts, transportation research generally attempts to develop indices and metrics with only limited attention toward policymaking and community-building (Chen 2019; Guerra, Cervero, and Al 2017). Both my qualitative and quantitative research for this book is not only located outside of the global north but also draws on a decolonial approach to transportation planning (see Chapter 2). That means decentering the dominance of northern-based motorized transportation-based research and instead focusing on more egalitarian modalities such as walking. Indeed, by considering how walkability has been applied in African cities, this chapter problematizes the overarching discourses around walkability in general.

Walkability in African Cities

Walkability is not a new feature of cities. In the Middle Ages, European cities were by necessity walkable, with an average spread of just a half-mile; and many of these cities remained walkable through the industrial era. Urbino, Italy, for instance, accommodated 30,000 people within an area of 300 acres, warranting a density of dwellings and diversity of overlapping activities; meanwhile, until the 19th century, every point in Boston, United States, was less than one mile apart (Southworth 2005). The introduction of transportation services in the 20th century – the horse-drawn cart, the horse-drawn tram, the electric tram, and the automobile – fueled the dispersal of urban life and the accompanying decline in walkability (Alca˘ntara De Vasconcellos 2004).

Walking, however, remains central to the economic and social life of African cities – even in those cities that are not particularly walkable. In Africa, car ownership is low – with just 3 percent of the world's registered cars (United Nations Environment Program and United Nations Human Settlements Program 2022) – while reliance on non-motorized modes like walking is higher, with on average 50 percent of trips in African cities being made on foot (Pendakur 2005). At the same time, Africa has 20 percent of worldwide road traffic deaths, and of the 260,000 people killed on African roads in 2019, 36 percent were pedestrians (United Nations Environment Program and United Nations Human Settlements Program 2022). Moreover, there tends to be an incongruity between urban design and walkability, with few cities adequately planning for walkability or providing appropriate infrastructure. Instead, communities work together to build and develop pathways, to develop shortcuts across the grass rather than taking the longer paved route and build pathways around obstacles.

Despite the regularity of walking in Africa, just 19 of the 54 countries have a walking policy. Planning requires the support of a range of departments and sectors – from transportation and urban planning to education, finance, and health. As a result, the majority of African cities lack a planning tool to support and sustain walkability. In turn, policymakers devalue walking and its economic benefits (Benton et al. 2023). South Africa by contrast – along with Ethiopia, Kenya, Nigeria, and Uganda – has developed a comprehensive policy framework for non-motorized transportation that references "economic vitality," "improved public health," and "greater social cohesion" as some of its goals (United Nations Environment Program and United Nations Human Settlements Program 2022).

The literature on walking and walkability in African cities is likewise scarce. When research on non-motorized trips has been conducted, it often only analyzes walking where it serves the main mode of travel rather than as part of a longer trip; and in other instances, walking is inaccurately combined with cycling into a single category of research (Behrens 2005). Recently, however, a number of studies of walking in Africa have tried to shift the discourse. Studies of walking in Accra (Okyere et al. 2023), Cape Town (Behrens and Makajuma 2017), Dar es Salaam (Andreasen and Møller-Jensen 2017; Masaoe 2017), Freetown (Oviedo et al. 2021), Johannesburg (Wood 2022b), Nairobi (Mitullah and Opiyo 2017; Nyamai 2021), and Praia (Anciaes, Nascimento, and Silva 2017) as well as across Africa (Benton et al. 2023; Mitullah, Vanderschuren, and Khayesi 2017) – have presented the details of walking as well as the associated planning processes and institutions. To Oviedo and others (2021), the literature of walking in Africa is generally studied in reference to three factors: safety from oncoming traffic and poor planning (Behrens 2005), health and its potential to alleviate non-communicable diseases (Oyeyemi et al. 2013), and unequal access to economic and social opportunities (Anciaes, Nascimento, and Silva 2017).

From another perspective, Benton et al. (2023) note three main challenges for walkability in African cities – the incongruity between walkability and economic development, the social stigma of pedestrians, and overvaluation of automobility. In African cities, walking remains the preserve of the poor and those with any means drive. Walking stands in contrast to capitalist and colonial notions of wealth and status. Indeed, walking remains synonymous with poverty, the very antithesis of urbane and modern. As a result, many cities lack adequate walking infrastructure and few governments invest in policy or planning reforms. Those walking in Freetown, for instance, walk because of a lack of public transportation, high costs of transportation fares, and because walking is faster than other alternatives (Oviedo et al. 2021). Benton, et al. (2023) note a lack of capacity and lack of funding as a key barrier to expanding walkability. To reconcile these limitations, they suggest policymakers shift how they determine the success of walking infrastructure and policies, build better relationships between stakeholders, and a better association between top-down planning and outcomes.

For African cities, there are two additional areas often discounted in the walkability literature: what I call "the unpleasantness of walking" and "walkability-with-people." While the literature promotes the economic and social benefits of walking (Litman 2003), it insufficiently reflects on "the unpleasantness of walking" and the factors that lead would-be walkers to reject bipedalism. Likewise, much of the existing discussion considers the role of planners in developing plans for walkability (Manaugh and El-Geneidy 2011), overlooking "walkability-with-people" and the ways in which communities can and should be involved in designing and developing pedestrian infrastructure and systems. These are particularly important additions to the discussion because they enable us to follow an egalitarian perspective on transportation justice, which reasons that transportation has a social benefit for all users (and would-be users) and therefore investment should be made to increase access for the maximum number (Van Wee and Geurs 2011) (see Chapter 1).

Indeed, transportation scholars have insufficiently considered the difficulties of walking. Forsyth (2015), however, reflects on some elements of the unpleasantness of walking in her research. She sorts the literature into three categories: research that considers how to create walkability, research that assesses walkability, and research that uses "walkability as a proxy for better design." Within the first category, Forsyth makes a distinction between walkability as a synonym for "traversable" (e.g., the urban conditions that facilitate movement) and "compact" (e.g., short distances between destinations), "safe" (e.g., perceived to be without crime), and "physically enticing" (e.g., pedestrian amenities and points of interest). In the second category, Forsyth considers the sociability of the walkable environment, as well as environmental and health elements of walkability. The third category, which includes city-wide factors of livability and sustainability, considers what she calls "a holistic solution to improving urban areas" (2015, 3) and is the outcome of the first and second categories. These categorizations add an important distinction between walkability as a feature of compactness and walkability as a pathway. Achieving these aims requires different approaches to city planning, a theme that will be further considered in the case of Johannesburg.

Likewise, scant attention has been given to the people actually doing the walking. Lo (2009), however, suggests that it is essential to define "pedestrian" in order to fully understand walkability. The Oxford English Dictionary (2012) defines a pedestrian as "a person walking rather than travelling in a vehicle," whereas the Merriam-Webster Dictionary (2016) specifies those walking with purpose (e.g., commuters) or recreation (e.g., strollers). Underlying these definitions are two types of pedestrians: those who choose to walk for reasons relating to convenience, health, or sustainability and those who are stuck walking because of cost and need. The array of pedestrian movement can be further subdivided into four categories: access mode (i.e., walking as the primary mode of transportation), access sub-mode (i.e., walking as part of another mode of transportation), circulation/exchange (i.e., leisure walking or window-shopping), and recreation (i.e., exercise) (Pharoah and Buchan 1997). Since walking proffers a range of benefits for both the walker and the public including cost savings, health, and the efficient use of land and resources, people need to be at the center of planning for walkability. This chapter therefore considers the experience of pedestrians as well as their role in planning walkability and walking-as-transportation.

This chapter conceptualizes walkability not only by the extent to which the built environment supports walking but also based on broader attempts to address the unpleasantness of walking and walkability-with-people. The discussion that follows employs Forsyth's (2015) categorizations of the walkability literature as well as Lo's (2009) attention to the pedestrian, both of which demonstrate the need for a more robust reconceptualization of walkability. In the discussions that follow I apply this conceptualization of walkability to Johannesburg and the attempts by Johannesburg city planners and politicians to develop walkability policies and projects.

Walking-As-Transportation in Johannesburg

Walking has been integral to Johannesburg's urban development since the 19th century, when workers lived within walking distance of their place of employment. "As long as walking remained the principal mode of travel," explains Hart (1984, 155), "the town, by necessity, had to be small and densely populated with little separation between place of work and place of residence and Johannesburg conformed to this form in virtually all respects." Before the introduction of the electric streetcar in 1890, Johannesburg measured approximately 2 kilometers from east to west and 1.5 kilometers from north to south (Hart 1984, 155). It was therefore very compact and easily traversable by foot.

In the 20th century, Johannesburg's urban growth was fueled by the introduction of horse-drawn trams in 1891. From the beginning, transportation was exclusionary: horse-drawn trams were reserved for White prospectors wealthy enough to pay to be transported between their homes in Fordsburg and the mines in Jeppestown (Van Onselen 2001). When the electric tram was introduced in 1906, it further entrenched the separation of Blacks and Whites, rich and poor (Sey 2012). By the 1920s, however, patronage was in decline, and the trams that once drove the expansion of the city were now impeding its growth (Wood 2015a). The Spencer Commission of Enquiry of 1934, tasked with determining the viability of trams, concluded that the density in Johannesburg was too low to adequately support public transportation (see Chapter 3). It stands to reason that Johannesburg was also becoming less walkable by this time.

In postwar Johannesburg – as in most cities around the world – cars became the symbol of modernity and came to dominate the city. Private car ownership was heavily encouraged while walking beyond the neighborhood was discouraged. Engineering concepts like the "neighborhood unit" (Clarence Perry's planning practice of designing self-contained residential neighborhoods distinguished by high-capacity roadways, popular in American cities in the 1920s) and the "environmental area" (Colin Buchanan's practice of isolating urban activities from noise, pollution, and traffic, popular in British cities in the 1960s) became commonplace in apartheid South Africa. However, these models, according to Behrens (2005), were inappropriate in South African cities where walking generally takes place beyond the confines of a residential neighborhood or business precinct.

Moreover, during apartheid "roads became symbols of oppression and sites of resistance," writes Bremner (2004, 108–109). "Forced removals, the movement of troops and armored vehicles, marches, road blocks, running street battles: the culture of the street was a highly politicized one," she continues. Apartheid-era town planners gave little thought to walking or walkability, and for the next few decades, planners were focused on building asphalt highways to accommodate the White elite who could afford to drive (Wood 2020a). Unsurprisingly, apartheid-era planners only gathered data for motorized trips and little is known about walking during this era (Behrens 2005).

In the postapartheid era, "the road was liberated" from serving as a symbol of domination, explains Bremner (2004, 108–109). The road was then "claimed by a myriad of small time traders, domestic workers and informal institutions, as a site of conviviality, livelihood and leisure." City planners responded to this liberation by supporting alternative approaches to transportation, one of which was making Johannesburg more walkable. Unfortunately, these attempts have been marred by a muddled interpretation of the concept. Just as the scholarship delineates between understandings of walkability as a synonym for "compact" and as a synonym for "traversable" (Forsyth 2015), there is likewise a disconnect between policies that aim to make Johannesburg more compact and projects that aim to make it easier to traverse.

Today, walking infrastructure in Johannesburg is representative of broader economic and social inequalities: whereas some areas of the wealthy northern suburbs are characterized by a nearly ubiquitous lack of sidewalks, other areas have extremely wide sidewalks (See Figure 5.1). Likewise, some neighborhoods

Figure 5.1 Pedestrians walk in blocked cycle lanes in Hillbrow, Johannesburg (photograph by author, February 2020). One of the many challenges for walking in South Africa is insufficient space on the sidewalks.

are defined by high-density developments containing a mixture of residential and retail in proximity to public transportation, while others exhibit low-density urban sprawl. Cabaret (2012) depicts Johannesburg as being composed of a highly differentiated set of streets – private streets, village main streets, township streets, downtown streets, and fake streets – in which contradictory norms are implicitly applied. Across the city, vastly dissimilar urban conditions encourage or discourage walkability.

The profile and practices of the pedestrian are equally diverse: while some Johannesburg residents who typically drive would prefer to walk but say they are afraid, many others dream of owning a car instead of walking (Gaylard 2017). Age, gender, class, and economic status are all reflected in this unequal access to mobility (Cabaret 2012). Because of these vast inequalities, Johannesburg has a disproportionately high number of walkers comprising primarily commuters marginalized from automobile usage and ownership. Walking infrastructure in Johannesburg therefore mirrors the social tensions and spatial inequality of the postapartheid city (McCool 2016).

The Failings of Walkability in Johannesburg

Attempts to improve walkability in Johannesburg overlap with efforts to redress the economic and social inequalities left over from decades of apartheid spatial planning. The City of Johannesburg launched a series of plans to promote walking and street-level interactions: the Framework for Non-motorized Transport (City of Johannesburg 2009), the Complete Streets Design Guidelines (City of Johannesburg 2013b), and the city's Strategic Integrated Transport Framework (City of Johannesburg 2013a). These reports categorize walking as both an access mode and an access sub-mode where appropriate. They call for new infrastructure to support walking as well as better integration with public transportation. At the broadest level, the 2013 Strategic Integrated Transport Framework calls for transportation infrastructure to improve the quality of life for current and future generations; while in a more prescriptive sense, Johannesburg's Non-Motorized Transport Framework identifies walking as the preferred mode for short distances, with student-school trips not exceeding 30 minutes and convenience-related trips to shops and services of up to 15 minutes. It also requires all new developments to incorporate sidewalks through the site and immediately adjacent to the site boundary. These documents are backed by the 2040 Growth and Development Strategy (Joburg 2040), the primary document guiding Johannesburg's planning, which calls for greater social inclusivity and cohesion, reasoning that improvements to transportation and mobilities will further these societal goals.

Johannesburg's Corridors of Freedom project is its most ambitious approach to walkability. The plan aims to make the city more compact through a series of interventions that combine new transportation infrastructure with high-density, mixed-residential, retail, business, educational, and leisure facilities (see Chapter 4). And the plan acknowledges a central role for walking and walkability to transform the economic and social fabric of Johannesburg. Non-motorized transport "is a crucial part of the transport system," contends the Strategic Area Framework for the Empire-Perth Development Corridor, and "everyone should have access to urban opportunities and mobility with the necessary safety precautions in place" (City of Johannesburg 2012a, 26). In addition to upgrading parks, clinics, libraries, and sports facilities, the Corridors of Freedom project plans for pedestrian bridges across highways and railway lines. These walkways were selected for their potential to connect rich and poor, White and Black, residential and commercial areas in adherence with the 2013 Complete Streets Design Guidelines. The following section discusses the construction and development of these pedestrian bridges and their attempts to make Johannesburg more walkable.

In support of these walkability policies, a variety of pedestrian projects were introduced: sidewalks and crosswalks (JDA 2018a); pedestrian-oriented land use and building design (JDA 2018b); increased road connectivity with additional shortcuts for pedestrians and cyclists (JDA 2020); traffic calming measures and vehicle restrictions (JDA 2018a); safety education and law enforcement (JDA 2018b); and vehicle parking policies and fuel taxes (JDA 2018b). There were also less costly investments in trees and shrubs, lighting, benches, and security cameras (JDA 2015b). However, by far the most elaborate, expensive, and evident result of postapartheid investment in walkability is the pedestrian bridges built in Johannesburg. For Johannesburg planners, the success of walking projects is associated with an increase in the number of pedestrians, pedestrian safety, public transportation ridership, and economic activity [16; 18; 19]. To date, however, none of the pedestrian bridges have reached their goals. The remainder of this section analyzes three pedestrian bridges – the Grayston Pedestrian Bridge, the Milpark Pedestrian Bridge, and the Westbury Pedestrian Bridge – including their failings (see Figure 5.2).

"A Very Expensive Way to Deliver an Otherwise Cheap Mode of Transportation"

The first example, the Grayston Pedestrian and Cycle Bridge (see Figures 5.2 and 5.3), which opened in March 2018 and forms part of the Great Walk, is a multimillion-Rand high-profile investment to support walking between Alexandra (a historically black area and now a very deprived neighborhood) and Sandton (Johannesburg's secondary CBD and Africa's wealthiest square mile, home to international companies and the city's wealthy elite). The Great Walk is a pseudonym for the 5.2-kilometer journey made by an estimated 10,000 domestic workers, gardeners, and shop workers on foot across Johannesburg's busiest highway, the M1, every morning and evening. The bridge is expected to have far-reaching impact by replacing a previously a dangerous and difficult trek, with people scurrying along the concrete shoulder of a car bridge and/or walking circuitous routes to avoid vehicular traffic (JDA 2015a). The Grayston Bridge is not only meant to provide a new pedestrian experience but also to transform the profile and practices

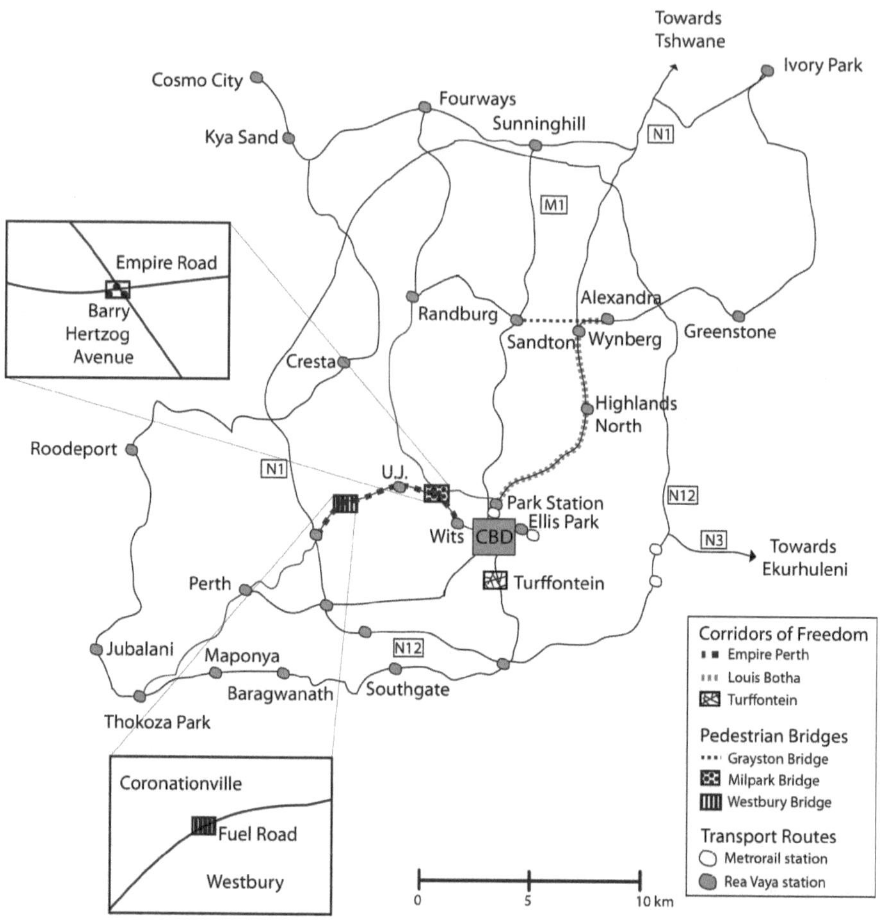

Figure 5.2 Map of Johannesburg's Corridors of Freedom and pedestrian bridges (map by author, August 2021). These three pedestrian bridges – the Grayston Pedestrian Bridge, the Milpark Pedestrian Bridge, the Westbury Pedestrian Bridge – were integral to the planning for Johannesburg's three Corridors of Freedom.

of walking in both the wealthy and deprived parts of the city. The bridge, according to Christine Walters, then-Mayoral Committee for Transport, "literally bridges the divide between these two contrasting communities and creates a visual gateway into the economic heartbeat of our city" (City of Johannesburg 2015). Likewise, the Graystone Bridge serves as a synonym for good transportation planning. Sello Lemao, then-Member of the Mayoral Committee for Public Safety, concluded that transportation interventions such as the pedestrian bridge served to transform the city by making it more "sustainable, livable and economically vibrant" (City of Johannesburg 2015).

Figure 5.3 The Grayston Pedestrian and Cycle Bridge (photograph by author, February 2020). Johannesburg's Grayson Pedestrian and Cycle Bridge, which opened in March 2018 and forms part of the Great Walk, is a high-profile investment to support walking between Alexandra and Sandton.

In May 2019, the City conducted an assessment of non-motorized transportation on Grayston Drive to determine the success of the pedestrian bridge. Of the more than 5,000 pedestrians crossing the M1 highway on Grayston Drive, only 26 percent used the pedestrian bridge. The remaining majority of pedestrians continued to walk on the original footpath on the southern side of the vehicle lanes on Grayston Drive (Loots 2019). The report goes on to suggest that widening existing sidewalks on either side of the vehicle bridge "might have been a more cost-effective way of solving the problem" (Loots 2019, 6). Moreover, such an approach would have better considered "the people who need to cross the bridge between Alexandra and Sandton," the report concludes (Loots 2019, 6).

By those key community-led metrics, the Grayston Bridge is a failure. Pedestrian usage, pedestrian safety, public transportation ridership, and economic activity are all below anticipated levels (Loots 2019). Moreover, with increased costs, long delays, and widely publicized accidents, critics question this expensive investment in pedestrian infrastructure (Brown 2015). For the most part, the wealthy residents of Sandton still frown upon pedestrianization while poorer Alexandra remains chronically underinvested (Kekana and Allison 2019). Pedestrian bridges are "flashy," explains a transportation consultant in Johannesburg, "but without changing people's approach to walking … so it becomes a very expensive way to deliver an otherwise cheap mode of transport" [18]. Another expert recalled his hopes that the bridge would shift how and where Johannesburg residents shop:

"We had imagined that with more people walking there would be more retail activity on the ground-floor level, but none of that has happened" [24]. Critics go further, alluding to the unpleasantness of walking by calling it preposterous to presume that a pedestrian bridge would improve the experience of pedestrians when the vast majority of the pathways remain unpaved and unnoticed [26]. Likewise, projects like the Grayston Bridge were not designed with the pedestrian in mind. The next two sections go into more detail about the unpleasantness of walking and walkability-with-people.

The Unpleasantness of Walking

The Milpark Pedestrian Bridge, completed in December 2019, provides an example of how the City has misunderstood who is actually using the pedestrian bridges. The Milpark Bridge sits at the intersection of Barry Hertzog Avenue and Empire Road (two four-lane arterials), in an area that incorporates a mixture of land uses, including business, retail, residential, education, and health facilities. High traffic volumes on both Barry Hertzog Avenue and Empire Road were seen as obstacles to pedestrian movement, creating disconnections between institutions and landmarks. Rather than focusing on the underprivileged pedestrian, the bridge was conceptualized to attract high-end commercial and residential density to the area (JDA 2020). It also targeted students traveling along the Rea Vaya bus rapid transit route to the two universities (University of Witwatersrand and University of Johannesburg) [26]. As a result, opportunities were missed to introduce a diversity of low-scale businesses such as fruit sellers, newspaper stands, and snack shops [24]. Existing pedestrians and their daily needs were insufficiently considered, and the resulting bridge therefore not only fails to attract additional walkers but also fails to provide an enhanced experience for existing pedestrians.

The Milpark Pedestrian Bridge provides an opportunity to reflect on the extent to which walking is thought to be a universal good with many benefits (Litman 2015; Southworth 2005): it is cost-effective, flexible to changing economic and social opportunities, healthy for pedestrians, and sustainable for the environment (Ewing and Cervero 2010). There is, however, a noteworthy disagreeableness to walking – it is time-consuming, slow, uncomfortable, and at times dangerous for the pedestrians, and it is contingent on the physicality of the pedestrian as well as the urban conditions available. Some of these challenges (e.g., the urban conditions) can be overcome by policy and infrastructure, but others (e.g., physicality of walking) cannot. People walk in Johannesburg because they are "transport poor." According to Lucas (2018), a person is transportation poor if any of the following conditions apply: the transportation options do not reach either the origin and/or the destination; the available transportation modes are not suited to the person's physical capabilities; the cost of transportation pushes the household below the poverty line or the duration of the journey is excessive; or the conditions of the journey are unsafe or otherwise unhealthy. Under any of these circumstances, a person's quality of life, health, or opportunities for improvement will be restricted. In Johannesburg, "people walk extremely long distances to avoid paying for transport" [1].

Low-income earners spend more than 25 percent of their monthly earnings on transportation, and high unemployment further limits modal choice (Ratshilumela 2018). Consequently, walkability should have been motivated by a pro-poor agenda that recognizes walking not only as the last mile of motorized transportation but as a complete transportation option. Likewise, a pro-poor agenda recognizes the value of walking as a resource available to all people. Policies and projects in Johannesburg are unfortunately inadequately aligned with this understanding of walkability and therefore fail to cater adequately to the pedestrian experience.

Walkability-with-People

The Westbury Pedestrian Bridge and Park provides an opportunity to reflect on how walkability projects can fuel, and be fueled by, partnerships between the city and the community. Situated between two deprived communities, Coronationville and Westbury, the infrastructure literally aims to bridge a divide (see Figure 5.2). Interestingly, the articulation of the need for this pedestrian bridge came from the Westbury community itself, who during a stakeholder consultation with the city requested a pedestrian bridge for people to move safely across Fuel Road. In 2013, the City surveyed 6,300 households across the city, asking them about their transportation movements and levels of satisfaction. Issues of last-mile accessibility were identified. The City then held workshops asking stakeholders – councilors, transportation operators, residents associations, and commuter groups – to outline their transportation challenges and their views on the proposed plans (Transport Department 2016). The community expressed concern that the existing on-grade crossing was dangerous not only for adult pedestrians but also for school children. They also complained that the new Rea Vaya station was difficult to access without an overhead connection. In response to these stakeholder engagements, the City then commissioned a study that revealed that over 500 people per hour cross the road during the morning and evening peak times (JDA 2016). The resulting bridge, however, fails to incorporate much of what the community wanted: it includes a park, for instance, which cost R25 million but is without adequate security, and this space turns into a no-go area after dark; moreover, it is particularly dangerous for school children during the day. The community requested a safe pathway across the road but by replacing one problem – the dangers of traffic – the City facilitated another safety concern – crime. This is an example of a missed opportunity to collaborate on the design and development of a walkability project in Johannesburg.

One of the main reasons why the Westbury pedestrian bridge fails is because city officials did not develop schemes that encourage walkability-with-people. Johannesburg's policies tend to presume residents to be the users rather than the drivers of better urban planning [2]. Accordingly, the City assumes a weighty role for urban planners in guiding walkability [14] and while it is certainly important for cities to put forward frameworks and policies, the agenda should be built in partnership with local communities. Indeed, there is an opportunity as well as a need to develop policy alternatives "that maximize the minimum level of primary goods to the people in the worst-off position" (Pereira, Schwanen, and Banister 2017, 175) so that "everybody should be well-off up to a certain minimum threshold" (Lucas, Van Wee, and Maat 2016, 477). This would fulfill a transportation-justice mission (Rigolon, Toker, and Gasparian 2018), as well as enable cities to develop transportation solutions that actually work for the community utilizing them.

Conclusion

An African flâneur, as depicted in Ivan Vladislavić's novel *Portrait with Keys* (2006), is "anxiously alert, but nevertheless manages to immerse himself in the city" (Gaylard 2017, 57). For many, "Johannesburg discourages flâneurism," writes Gaylard (2017, 60); indeed, "the people who walk Johannesburg daily are not flâneurs at all, but migrants, or workers, to whom the city still denies the right to public transport" (Gevisser 2014, 20). "It takes a person of some physical stature and courage to roam its streets," continues Gaylard (2017, 60), and those "who pound the pavements on a daily basis do not do so with impunity and without trepidation." I introduce the concept of an "African flâneur" to draw attention to the important role of the pedestrian, rather than the infrastructure, in walkability projects. Too much attention is paid to the glitzy infrastructure in Johannesburg, and not enough attention is paid to the people actually using it. As a result, much of it is under-utilized.

The concept of the "African flâneur" is also relevant to transportation planning. Rather than evidencing the merits of walkability or the ease of its application, this chapter highlights the challenges of actually doing walkability. First, by contemplating the varied definitions of walkability, this chapter reveals an ambiguity between the literature that interprets it as a synonym for compactness or a substitute for traversability. Next, by considering how walkability has been experienced unevenly in African cities. I then apply the concept of walkability to Johannesburg by examining policies that promote compactness and infrastructure projects that enable traversability. While the municipality uses walkability as a synonym for livability and sustainability, without a clear focus on the pedestrian these concepts prove difficult to apply. This leads to a discussion of the shortcomings of Johannesburg's planning to explain why walkability has failed to take root. This is not to dwell on the failures in Johannesburg, but rather to reflect on the limitations of the focus on walkability as a top-down policy and the challenges that emerge in its application.

This chapter furthers our discussion of why transportation fails, first by providing an example of urban planning in an African/global south city, and second by attempting to decolonize transportation planning by highlighting Johannesburg's experiences. A decolonized approach to transportation planning brings pedestrians and pedestrianization to the forefront (see Chapter 2). Rather than being seen as a marginal modality or a last-mile alternative, the logic of decolonizing turns walking into a primary mode of transportation. Just as the settler-colonials subsumed economic, cultural, and social life, they also created a hegemony of car usage. Decolonizing therefore means restoring power to pedestrians. The call to decolonize is particularly salient in South Africa, where transportation has historically been used to control and divide cities. This study of walkability thus fits within ongoing efforts to remedy inequalities in transportation, by elevating the role of pedestrians and calling for better infrastructure planning and community consultation to support pedestrianization. In reflecting on the concept of walkability, I hope that city planners in Johannesburg and beyond will design and develop spaces that encourage and enhance walking for all. This discussion continues in Chapter 6 with a focus on trans-paw-tation.

6 Trans-paw-tation

Introduction

Animal geographies constitutes a burgeoning subfield in geography. The discourse engages with a broad array of epistemological and methodological scholarship from anthropology, cultural theory, politics, and sociology as well as geography to highlight the agency of animals. Studies of bats (Thomson 2007), bison (Lulka 2008), chickens (Hovorka 2008), dogs (Fletcher and Platt 2018; Urbanik and Morgan 2013), donkeys (Hanekom 2004), elephants (Lorimer 2010), monkeys (Yeo and Neo 2010), and pigeons (Jerolmack 2008) – to name just a few – dislocate the anthropocentric dominance of geography and blur the binaries between human and non-human, domestic and wild, and tactile and ethereal. This chapter aims to further complicate the discourse by adding a focus on mobilities suggesting additional ways in which animal geographies can enrich this discussion of transportation failure.

This chapter aims to shift away from anthropocentric approaches that understand transportation failure within traditional human and urban forms. Instead, it highlights "trans-paw-tation" in South African cities and the theoretical possibilities of thinking across animal geographies and mobilities. In so doing, this chapter aims to bridge the lacuna between animal geographies and African studies. Scholars of animal geographies have begun to challenge this dissonance – for instance, Duffy and Moore's (2010) study of neoliberal elephants in Botswana, McGregor's (2005) analysis of conservation in Zimbabwe and Massé and Lunstrum's (2016) analysis of poaching in Mozambique – but there is both opportunity and need to rethink Africa from the perspective of animal geographies. This approach contributes a layer of complexity to the conversations in animal geographies as well as the interdisciplinarity of transportation scholarship.

Second, the chapter proposes to bring animal geographies into conversation with urban studies. Indeed, just as animals have been systematically excluded from modern cities, so have they been overlooked by the urban studies discourse. Recent scholarship in animal geographies has begun to shift with a renewed attention to animals in cities. Wolch (1995, 2002), for example, has been the most explicit in attempts to depict animal geographies as urban. Locating these studies of animals in African cities is also increasingly important – Hovorka's (2008) chickens,

DOI: 10.4324/9780429322280-6

Wilson's (2018) livestock, and Rink's (2020) amaphela taxis demonstrate the possibilities of locating research at the intersection of animal and urban geographies in African contexts. Additional theorization of urban animals, in particular in global south and African cities, is warranted.

Third, this chapter reflects on methodologies for trans-paw-tation. Tactics for studying animal geographies are varied but research oftentimes includes a candid reflection on our methodologies (Buller 2015; Gibbs 2020). This chapter is written by a human researcher with only minimal contributions from animals (i.e., fiddo at my feet). However, it draws on critical animal geographies (Gillespie and Collard 2015), which are attentive to the active opposition to the ethically unacceptable material, instrumental, and conceptual treatment of animals by humans. My approach therefore relies on both historical and contemporary sources to develop a keener understanding of transportation in urban South Africa. These materials are categorized and controlled by humans; however, in exposing the central role of animal mobilities this chapter furthers the epistemic methodologies of animal geographies and transportation geographies.

The next section of this chapter pinpoints the origins of animal geographies and highlights today's dominant approaches. From there, this chapter takes an urban paws, focusing on the connections and disconnections between animal geographies and urban geographies. This chapter then outlines an approach to animal mobilities that draws across both animal geographies and mobilities scholarship. This leads to this chapter's three approaches for understanding trans-paw-tation: first by studying animal movements, second by considering animal legislation and policies, and third by examining the symbolism and metaphors that a combined understanding of animal geographies and mobilities furthers. These analytical reflections are then applied in urban South African cities through a series of historical and contemporary reflections. In the conclusion, it will be clear how we can use animal mobilities to better understand transportation failure in South Africa and beyond.

Finding the Animals in Human Geography

Attention to animal geographies has ebbed and flowed over the past 100 years: in the first half of the 20th century, according to Wolch (2002), geographers adopted either a zoogeographic perspective to highlight the spatial patterns and social relations between animals and humans, or a culturally-oriented animal lens that underscored biological and environmental concerns. In this "first wave" of animal geographies (Hovorka 2018; Urbanik 2012), zoogeographers – typically zoologists, ecologists, and population biologists – studied animal behaviors, habitats, and species-specific geographies, and they drew on empirical methods to understand how human actions impacted animals. For example, Mobius (1894) was concerned with the habits of whales, and Clarke (1896) noted shifting bird migration in the British Isles. The term "animal geography" is generally attributed to Newbigin (1913) and her book of the same name that called for geographical scholarship that incorporated biology and nature. While this scholarship drew attention to the intersections between animals and humans, it for the most part, disregarded how human action was harming animal life.

In the mid-20th century, during the "second wave" of animal geographies, cultural geography became more prominent (Hovorka 2018; Urbanik 2012). Geographers studied the associations between animals and landscape highlighting the centrality of animals in cultural studies – for example, animal domestication, animal husbandry, and the role of animals in agrarian societies (Wolch 2002). Bennett (1960) proposed a "cultural animal geography" to study animal–human relations and how humans shape animal life in much the same way that zoogeography spotlights the behaviors and experiences of the animals themselves. This scholarship is criticized for failing to acknowledge the agency of animals, instead treating them as "cultural artifacts" – "there to be dominated and used, or symbols of human culture and meaning" (Wolch 2002, 724). Beck's (1973) study of dogs in Baltimore for instance was not motivated by concern for the dogs but rather by the need to understand the effects of urbanization on humans.

In the 1990s, animal geographies once again surged in popularity in a "third wave" of scholarship (Hovorka 2018; Urbanik 2012). Two Special Issues – Wolch and Emel's (1995) "Bringing Animals Back In," for *Environment and Planning D* and Philo and Wolch's (1998) "Space, Place and Society–Animal Relations" for *Society and Animals* – and two edited collections (Philo and Wilbert 2000; Wolch and Emel 1998) signify the dawn of the "animal moment" (Wolch and Emel 1998) or the "animal turn" (Buller 2014) in human geography. Wolch (2002) attributes the expansion of animal geographies to the environmental movements starting in the 1970s and their lobbyist organizations such as People for the Ethical Treatment of Animals (PETA) and the Animal Liberation Front as well as intellectual interest in racism, sexism, and speciesism. For Philo and Wilbert (2000, 4), this moment of animal geographies "focus[es] squarely on the complex entanglements of human–animal relations with space, place, location, environment and landscape." This scholarship emphasizes animals as subjects of, rather than objects of, human manipulation. Accordingly, this approach enabled animal geographies to reach across human geography's sub-disciplines and build synergies with feminist, historical, social, and urban geographies (Buller 2014). The subfield reached its pinnacle when in 2009, Ogra and Urbanik founded the Animal Geographies Specialty Group of the American Association of Geographers (Urbanik 2012).

We are currently in a "fourth wave," with studies of bats (Thomson 2007), bison (Lulka 2008), chickens (Hovorka 2008), dogs (Fletcher and Platt 2018; Urbanik and Morgan 2013), donkeys (Hanekom 2004), elephants (Lorimer 2010), monkeys (Yeo and Neo 2010), and pigeons (Jerolmack 2008), increasingly commonplace in geography. Three series of papers in *Progress of Human Geography* demonstrate the continued effervescence of animal geographies, first by Buller discussing the ontological origins (2014), methodological approaches (2015), and ethical questions of animal geographies (2016); second by Hovorka examining the globalization (2017), hybridity (2018), and power relations of animal geographies (2019); and third by Gibbs connecting animal geographies with more-than-human geographies (2020) and "killing and caring" for animals (2021). For these scholars, "animals matter individually and collectively, materially and semiotically, metaphorically and politically, rationally and affectively" (Buller 2014, 310). They recognize the agency of animals as co-inhabitants, disrupt the dualisms of animal–human configurations, and establish a politics that accommodates this complexity.

In this fourth wave, animal geographers are increasingly engaging with a multiplicity of geographical approaches. Animal geographies scholarship aims to blur the dualities of human and non-human, artificial and natural, domesticated and wild, instead focusing on methods for repositioning animals beyond "conceptual othering" – "by setting them apart from us in terms of character traits" and a "geographical othering" – "by fixing them in worldly places and spaces different from those that humans tend to occupy" (Hovorka 2008, 97). This latest approach not only provides tremendous insight into animal-human relations but also overlaps with broader critical discussions of capitalism (Turner 2009), neoliberalism (Duffy and Moore 2010; Moore 2011), colonialism, postcolonialism (McGregor 2005; Nyman 2003) and decolonialism (Hovorka 2017). Analyses of inter-species (animal–animal) power relations, including differences between species as well as within species are also important (Hovorka 2008, 2019). For Hovorka (2018), the "hybridity" between animal geographies and other geographies enriches the discipline as a whole. Critical engagements with political ecology, for example, allow animal geographers to reflect on environmental issues, while political ecologists can think beyond animals as incidental to environmental conditions. There is also conceptual scope to bring animal geographies into conversation with actor-network theory, science and technology studies, and materiality.

Animal geographies is no longer "too removed from central problems of geography" as had previously been the case (Hovorka 2008, 111). However, despite so much scholarship, geography still has a long way to go to "hear the cry" of the animal (Gibbs 2020). There remains no singular approach to studying animal geographies but rather a range of epistemological and methodological approaches, all of which aim to better understand animals and their role in the world. This chapter adds trans-paw-tation as an additional approach that has been under-explored in the literature.

An Urban Paws

Animals are denizens of our cities: there are 500,000 dogs living in New York City (human population 8.5 million), as many as 1 million cats in Istanbul (human population 15 million) and more than 550 bird species in Hong Kong (human population 7.5 million). There are 240 pet stores and 172 pet daycares in London, and the Underground has been crowned the most dog-friendly public transportation system in the world.

Animals were first introduced into cities to facilitate growth and exploitation: horses pulled people and products across the growing metropolis, while livestock fed the ballooning population. In the 19th century, Chicago was synonymous with meat-packing (Philol 1995), leather was the second largest industry in Britain (Atkins 2012), and Australian urbanization is "indebted to the labor of camels" (Owens and Wolch 2017, 545). But American and European cities were cleansed in the 20th century and the uncouth sights and smells of livestock, stables, and wet markets were subsequently removed (Philol 1995). Even so, chickens and livestock continue to feed our cities (Hovorka 2008; Wilson 2018) and dogs and horses serve in municipal border and police forces (Owens and Wolch 2017). In the contemporary city, urban animals are characterized either as "tools","immigrants", and invasive (Shingne 2022). Shingne's (2022) categorization of animals as immigrants reflects this historical and contemporary rejection of animals in cities. In the 19th century, dogs are said to have invaded the urban space, while the expulsion of cats in New York City is associated with the removal of Chinese immigrants in California (Brady 2012). Today, urban animals like pigeons are often dubbed "flying rats" (Jerolmack 2008) and raccoons are nicknamed "trash pandas" (Shingne 2022). These metaphors highlight a widespread fear and condemnation of animals in cities – a "marginal social group" (Barua and Sinha 2019, 1171) – even where urban growth has encroached on animal habitats.

While animal geographies scholarship today takes special note of spatial and urban thinking, consideration for animals in urban scholarship has not always taken center stage. The Chicago School of Urban Sociology for example, despite their explicit ecological framing, overlooked the importance of animals in the city (Wolch 2002). Philo and Wilbert (2000) are perhaps the first to explicitly designate "animal spaces" as the "proper places stipulated for [animals] by humans" (2000, 13) from "beastly places," where animals are "getting on with their own lives and worlds without anything to do with us humans, performing their specific forms of agency to one another, creating their own worlds, their own beastly places, without reference to us" (2000, 19). The more recent scholarship has tried to locate animals within cities (Philol 1995; Wolch 2002) by drawing on an "urban theory that takes animals seriously" (Wolch 1998, 120). However, studies of "beastly places" are still lacking in the literature, mostly due to the methodological challenges of observing animal behavior in natural conditions in cities (Barua and Sinha 2019). In an effort to appreciate the spatiality of animal geographies, the scholarship considering "animal spaces" focuses on dogs (Urbanik and Morgan 2013) and chickens (Hovorka 2008), while those considering "beastly places" attend to pigeons (Jerolmack 2008), free-roaming cats (Wolf and Hamilton 2022) and monkeys (Barua and Sinha 2019; Yeo and Neo 2010).

Concepts such as "zoocities" (Holmberg 2015), "zoopolis" (Wolch 1998) and "transspecies urban theory" (Hovorka 2008) bring urban concerns into animal geographies. For Wolch, urbanization, by its very nature, necessitates the denaturalization of the environment and therefore the subjugation of fauna (and flora). And Hubbard and Brooks (2021) go on to argue that animals, just like low-income residents, can be victims of gentrification and that they have a right to remain in the city. Instead, animal presence should be considered in city planning and policymaking. Discussions of the zoopolis and transspecies urban theory therefore expose the "dialectical relationship between people and animals" (Hovorka 2008, 97), a relationship that is fundamental to the growth and maturation of the city. Wolch's "anima urbis" (2002) suggests that animal geographies enhance urban geography by revealing how animals shape the character, identity, and subjectivity of urban dwellers as well as the urban landscape. Animal geographies scholarship goes so far as to say that the city itself is a product of transspecies relations with animals (and their relationships with humans) shaping the form and function of the city.

For African urban geographies, animals have played a smaller, albeit still important, role in the literature. Rink and Crow's (2021) study of the "humanimal" assemblage of horse and horse cart driver in Cape Town and Hovorka's (2008) study of chickens in Gaborone both note the importance of human and animal interactions. Both authors recognize the marginalized role of animals as opposed to their human counterparts. For Hovorka in particular, chickens have influenced the form and function of Gaborone, "shaping possibilities for how the city looks (e.g. through urban planning mechanisms), how the city functions (e.g. through local economic dynamics), and how human dwellers experience urban life (e.g. through socio-economic empowerment)" (2008, 107). She demonstrates that while livestock has been excluded from the city – regarded as wild, unclean, and unhygienic, especially in urban Africa – the city and urban life cannot be disentangled from its animal–human relations.

An urban paws therefore involves rethinking urban theory from the perspective of the animal. In spite of these overlaps between animal and urban geographies, additional theorization is still needed to better understand how animals and humans intersect and intertwine, as well as how animals can serve as urban actors in the zoopolis.

Bringing Animal Geographies into Transportation

Animals, domestic and wild, are constantly on the move: there's Boji, an Anatolian shepherd mix dog who lives on the streets of Istanbul and commutes by bus, ferry, tram, and train (Ayoubi 2021); and Moscow's stray dogs have learned to ride the labyrinth of the city's underground trains (Marquardt, Blakemore, and Eichenholz 2010). There's even a book about "Casper the Commuting Cat" (Finden 2011), written about a regular bus rider in Plymouth, England, and translated into six languages. Casper queues at the bus stop across from his house, boards the bus, curls up in a seat and travels 11 miles to the city center where he disembarks. Like their human counterparts, these animals are mobile, relying on mechanized transportation to move about. Indeed, there is much to learn by reflecting on mobilities and movement through the lens of animal geographies.

Animal mobilities scholarship is therefore attentive to the myriad of ways in which humans maneuver and manipulate animals: for example, walking dogs in dog parks (Urbanik and Morgan 2013) and keeping elephants in conservation sanctuaries (Duffy and Moore 2010). Some animals seem to be more mobile than others: dog-mobilities for instance have been studied historically (Brady 2012; Van Sittert and Swart 2003), in the urban imaginary (Urbanik and Morgan 2013), and in motion (Fletcher and Platt 2018), while elephant-mobilities are synonymous with conservation (Lorimer 2010), the tourism industry (Duffy and Moore 2010) and the ivory trade (Moore 2011). There have also been excellent attempts to think through the intersections between animal geographies and automobilities – for example, Flack's (2016) study of automobility in the English safari park. There are likewise efforts to understand animal stillness – for example, research into animal parks (Duffy and Moore 2010; Flack 2016) and zoos (Uddin 2015). All of these studies however still primarily engage with the effects of animal behaviors on humans.

The consequences of human actions on animal movement have been the subject of animal geographies from the beginning (Wolch and Emel 1995, 1998). Animal mobilities were inherent to the European conquest of the Americas: in 1493, Columbus took eight pigs to be the first domesticated animals in the New World, emblematic not only of colonial movements but also the conquistadors themselves – "hardy, mobile and physically adaptable" (Donkin 1985, 41). Van Sittert and Swart's (2003) history of dogs in South Africa notes the central role of the pre-colonial, colonial, and post-colonial canine. In colonial Africa, for example, "purebred dogs" were imported to serve as companions and security while "native dogs" were indiscriminately killed (Van Sittert and Swart 2003). By seeing animals as more than units of production or natural resources, animal mobilities provide an alternative framing through which to pinpoint inequalities in power and resources.

It is therefore important to challenge Western approaches to mobilities which privilege human movements over those of animals. Studies of animal mobilities are based in both the global north (Feldman 2009; Jerolmack 2008) and the global south (Hovorka 2008; Wilson 2018) – birds in Ghana (Pierre 2013), chickens in Botswana (Hovorka 2008), monkeys in Singapore (Yeo and Neo 2010), and jaguars in El Salvador (Campbell and Torres Alvarado 2011), all provide the scholarship with an international focus. Animal conservation and tourism drive much of this globalness. Duffy and Moore (2010) for instance draw on the case of elephant raising in Botswana and Thailand to highlight the "actually existing neoliberalism" within the tourism industry; and Massé and Lunstrum (2016) note a relationship between conservation, securitization, and capital accumulation in wildlife tourism in Mozambique. In recognizing their mobilities however, the scholarship risks further subjugation and repression.

In order to better understand animal mobilities, this chapter develops a framework for trans-paw-tation that draws on these epistemological and methodological approaches. First, it applies concepts from the mobilities paradigm – the politics, power, and meaning of movement promoted by the mobilities turn (Cresswell 2010; Sheller and Urry 2006) – to animals. It follows Hodgetts and Lorimer (2020), who consider not only how animal movements are shaped by humans but also how the concept, which is generally human-oriented, can be applied to the lived experiences of animals. They introduce the idea of "animals' mobilities" – "the embodied, affective, and lived animal experience of mobilities," as discrete from "animal mobilities" – "the movement (or stillness) of animals as shaped by the actions of various actors, particularly humans" (2020, 5). This distinction highlights animals' own mobilities rather than how animals have been placed and positioned by humans. Studies of animals' mobilities focus primarily on how an animal moves by considering its animal anatomy (e.g., paws, claws, and wings), diet (e.g., carnivore, herbivore), family structure (e.g., matriarchal), territories (e.g., migration or hibernation), breeding (e.g., mating-for-life or solitary), personality (e.g., brave or shy) or climate (e.g., cold or hot) (Hodgetts and Lorimer 2020). This first foray provides an important starting point for moving beyond the territorial fix of geography.

Second, this chapter considers how animal mobilities can also be understood through the policies and practices that rule and regulate movement and transportation. For Hodgetts and Lorimer (2020, 14), animal movements are controlled through Foucaldian notions of power and "the disproportionate ability of (some) humans to eradicate animals, or to achieve absolute control over their movements through technologies like the fence, the cage, the net, or the tank." While policies tend to exist in the human world, they also suggest an agency to animals that warrants legislation. The biopolitics of animal control of course originates from human-oriented power to move and the technologies that facilitate this control. That means looking not only at the dog walker but also at the leash through which power is distributed (Urbanik and Morgan 2013) or the horse cart rider as well as the bit, the harness, and even the cart itself (Rink and Crow 2021). This second approach therefore focuses on the machinations through which animal mobilities starts and stops.

Third, the chapter suggests that animal mobilities can be understood as symbols and metaphors for broader societal undertakings. Accordingly, animals provide an alternative lens through which to view transportation geographies. The greyhound as the symbol for the largest long-distance bus company in the United States symbolizes speediness; and the mynah bird on Durban's bus company represents reliability. The "camioneta" (meaning, chicken bus in Spanish), a derogatory term for over-subscribed urban–rural transportation in Guatemala, or the "gumba-kumba," referring to the high loads placed on top of the vehicle in Zimbabwe, is another way in which animals and mobilities intersect. Uncovering these disciplinary junctures provides tremendous insight into not only the aesthetic of the city but also the diversity of lifestyles and needs. Animal geographies have long been used to expose our societal flaws – zoos became synonymous with White, human, civilized conquest over the Black, animal, savagery (Uddin 2015); and there is the danger in discussing animals as "cultural artifacts" – "there to be dominated and used, or symbols of human culture and meaning" (Wolch 2002, 724). As long as the focus remains on the animals rather than the humans, there are boundless opportunities for this approach.

Finally, the chapter reflects on methodologies for studying animal mobilities. It discusses a range of mobile methods including "critter-cams," remote sensing, and GPS tracking (Hodgetts and Lorimer 2020). These methods aim to give animals a voice, rather than be spoken for (Buller 2015; Gibbs 2020). Animal movements however can be misleading or misinterpreted and it can be difficult to accurately interpret their decisions (Buller 2015). Animals have limited agency and cannot explain their movements, leaving the researcher to interpret their entanglements for them. Studies of dog owners (Fletcher and Platt 2018), opposed to the dogs walking, or poachers and/or conservationists (Massé and Lunstrum 2016), as opposed to the targeted animals, risk reifying the agency of humans over animals. Animal mobilities therefore also rely on ethnographic approaches to understand animal movements in both zoological (Duffy and Moore 2010; Flack 2016) and urban settings (Jerolmack 2008; Wolf and Hamilton 2022). Ethology, the study of animal behavior in their natural habitat, is particularly valuable for mobilities research (Barua and Sinha 2019).

Visual methods including digital mapping, film, and photography are increasingly being employed to think beyond "voice-based methodologies" (Bear, Wilkinson, and Holloway 2017, 227). For Bear, Wilkinson, and Holloway (2017, 227), visual methods help to decenter humans (and the researcher) by providing the animal with a means of expression. These methodologies, while limited, expose the movement and/or stasis of animals within an otherwise invisible landscape.

The remainder of this chapter applies this conceptual scaffolding to South African cities: first by studying the movement of animals and animals' movements, and specifically considering how animal-based transportation facilitated urban expansion. Second, the framework analyzes the policies governing animal mobilities, especially how animals are controlled within human cities. Third, the chapter explores the aesthetics of animal mobilities and how animals are represented in urban life. Together these three reflections provide insight into how to bring animal geographies into conversation with mobilities and in so doing realize the generative outcomes of failure.

Trans-paw-tation in South African Cities

Trans-paw-tation features prominently in South Africa, home to more than 200 species of mammals, 450 species of reptiles, 130 species of amphibians, 800 species of birds and 60 million humans. Animals are central to every sector – from the economy (e.g., lions and leopards appear on South African rand banknotes and 1.4 million tourists visit the Kruger National Park each year) to housing (e.g., the Red Ants Relocation and Eviction Services, or "Red Ants" for short, is a private security company noted for their red uniforms and their excessive use of force) to policing (e.g., the "Scorpions" were a special unit of the National Prosecuting Authority tasked with investigating organized crime and corruption which operated from 1999 and were replaced in 2008 by the "Hawks," a similar unit under the directorate of the South African Police Service). Transportation failure, I argue, is inherent to these animal geographies. The remainder of this chapter commits to exploring the intersections between animals geographies and transportation in South Africa: first, I will discuss the movement of animals and animals' movements. This section applies a historical lens to animal mobilities drawing out the associations between animals and South Africa's spatial development. Second, I will consider the policies that govern and control animal mobilities. The legislation discussed not only regulates the mobilities of animals but also where animals sleep, eat, and play. Third, I will highlight the broader aesthetic of animals' mobilities. Here the discussion turns to dogs and insects, and the ways in which their animal mobilities are politicized and prejudiced. These three tactics present an approach not only for connecting animal geographies and mobilities but also for thinking about the economic, political, and social uncertainties of urban life.

Movement of Animals and Animals' Movements

This first venture into animal mobilities is historical. Drawings of animal traction used by the Khoi-khoi and their herds of cattle and sheep are some of the earliest references to animal mobilities. Additional evidence of oxen used to transport goods reveals a history of animal power long before European colonization in 1652 (Joubert 1995).

Animal mobilities played a principal role in colonialism after the 17th century. Horses, donkeys, and oxen were used in agriculture (e.g., plowing the fields), and later in mining (e.g., hauling the deposits) and transportation (e.g., horse carts and horse-drawn trams). When Jan van Riebeeck landed in the Cape however, he brought no animals with him. Horse sickness and the difficulties of importing horses made it not financially feasible. Instead, the colonists relied on indigenous oxen and as such these animals played a role in the European colonization of South Africa (Joubert 1995). Horses were later imported from Indonesia, another holding of the Dutch East India Company, and this enabled the development of the coast, and later of the interior. The story repeats for donkeys, which were first imported to the Cape in 1656 (Hanekom 2004). And a further association of animal geographies with the colonization of South Africa occurred in 1897 when the Rinderpest plague – a European disease imported to South Africa by colonization – killed as many as 95 percent of cattle, leaving farmers and land owners destitute. Many were left with no option but to sell their labor for subsistence, which suggests a correlation between the loss of cattle and land and the expansion of the mines in South Africa (Jacobs 2001).

Animals are also deeply intertwined with urban development in South Africa. In Johannesburg, animal labor was first utilized to build the infrastructure and then move across it. Commissioner Street, Johannesburg's first road, was constructed by ox-drawn wagons filled with rocks and pulled up and down the street for a week. According to legend, "the roads (or wagon tracks) to Johannesburg were paved with skeletons of countless draught animals that died from hunger, thirst or exhaustion along the way" (Latilla 2013). Between 1891 and 1906, horse-drawn trams transported White miners between their homes in Fordsburg and the mines in Jeppestown (Van Onselen 2001). In 1906, Johannesburg replaced these horse-drawn streetcars with electric trams (Sey 2012). Horse-drawn transportation was seen as rudimentary and animal presence in the city archaic; their hoofprints were no longer acceptable within a modern metropolis.

Animal mobilities can also be explored by tracing the history of the urban animal. The donkey, for example, is indigenous to Africa – skeletons of the Nubian wild ass, which is generally believed to be the progenitor of the donkey, have been found in Egyptian tombs dating to the mid-4th century BC (Blench 2004). In the pre-colonial era, donkeys served as work animals, and in the colonial period, they were deeply involved in the construction and development of colonial cities. Donkeys carried loads, pulled carts and plows, and were used for riding and other forms of transportation. In contemporary South Africa, horses pulling horse carts are the last remaining urban work animal, with more than 250 working horse carts in the Cape Flats in Cape Town (de Klerk, Quan, and Grewar 2020). They generally work as collectors of scrap metal and garden refuse. Rink and Crow (2021) provide an account of the "humanimal" assemblage formed between horse and rider through their everyday mobilities. Tracing the history of the donkey and the horse in South

Africa exposes not only the movements of animals but also the economic and political geographies of everyday life.

This foray into the movements of animals is an important starting point for this framework for animal mobilities; however, much more can be learned by analyzing the legal geographies through which animals and humans coexist in urban South Africa.

Policies Governing Animal Mobilities

Another method by which animal geographies intertwine with mobilities is through governance and policymaking. This second discussion analyzes South Africa's animal protection laws and how these simultaneously empower and constrain mobilities. South Africa is a particularly remarkable context in which to reflect on animal policies because its progressive constitution does not include a clause protecting animals. Accordingly, animal sentience is not recognized in either legislation or policy. However, animal protection is intrinsic to South Africa. Laws governing animal protection date back to the Prevention of Cruelty Act in 1914 (South African National Government 1914). This legislation was followed by the Performing Animals Protection Act of 1935 (amended in 2016), which regulates "the exhibition and training of performing animals and use of dogs for safeguarding" and empowers a magistrate to issue licenses to "fit and proper persons" for exhibiting and training performing animals (South African National Government 1935). While these laws provide only minimal protections for the animals, their establishment recognizes the central role animals hold in cultural affairs in South Africa.

The Animal Protection Act No. 71 of 1962 (South African National Government 1962) demonstrates a firmer commitment to protecting animals from abuse. It expands the definition of animals to include "any equine, bovine, sheep, goat, pig, fowl, ostrich, dog, cat or other domestic animal or bird or any wild animal, wild bird or reptile which is in captivity or under the control of man" (South African National Government 1962). It also extends to include pets – "domestic animals and birds, and wild animals, birds, and reptiles that are in captivity or under the control of humans." The Act permits a maximum of two dogs per dwelling unit, three dogs in a freestanding house, four dogs in a property of more than 600 meters and six dogs on an agricultural property. For cats, no more than four cats are permitted in any residence and six on a farm. Likewise, the type of pet is regulated. Only dogs, cats, fish, birds or rodents are permitted by right and an additional permit is required to keep any other animal in a residential area. The Act outlaws any form of animal cruelty: pet owners must provide adequate water and shelter for their animals. These policies not only supervise animal movement but also establish the rules that govern domesticity of animals in South Africa.

Animal policies not only govern the movement of animals but also the stoppage and stillness – where animals eat, sleep, and play are important elements of animal mobilities. The Animal Protection Act of 1962 monitors the quality of animal dwellings in urban areas. Cruelty is defined by the Act as: "affording inadequate shelter, light or ventilation in which such animal is excessively exposed to heat, cold, weather, sun, rain, dust exhaust gases or noxious fumes." The Act specifies that "making adequate provision for suitable food, potable water and rest" for the animal is a minimum requirement for pet ownership. It explicitly disallows an animal to be confined by "chains, tethers or secures unnecessarily or in such a manner as to cause that animal unnecessary suffering." Accordingly, animal legislation not only governs how animals move but also when, where and for how long they stop.

This second venture into the policies that govern animal mobilities provides additional insight into the ways which animal movement is legislated by human and urban forces. What remains to be seen is how animals can expose a broader understanding of social life. It is the final discussion of animal imagery that illuminates the importance of thinking about animals mobilities in South Africa.

Imagery and Aesthetics of Animal Mobilities

Animal geographies provide a symbolic lens through which to understand mobilities. Animals can serve as representations for economic, political, and social mobilities as well as movement. In South Africa, animals are especially emblematic of colonialism and apartheid. Dogs and insects are just two of the many important specimens through which to understand oppression and marginalization.

Dogs are a frequently used lens for animal geographies (Fletcher and Platt 2018; Urbanik and Morgan 2013), especially in South Africa (Van Sittert and Swart 2003). This is partly because dogs have a long history in South Africa: San rock paintings of dogs date back to the Iron Age and dogs were present when Vasco da Gama landed at St Helena Bay in 1497 and famously remarked, "They have many dogs like those in Portugal, which bark as do these" (Boonzaier et al. 1996, 54). Van Sittert and Swart identify a pre-colonial, colonial, and post-colonial dog as well as "Africanis," a "dark doppelganger, the wild 'Kaffir' or stray dog" (2003, 139). The kaffir dog "combines the abusive racial insult 'kaffir' with the volatile meanings of 'dog'" and because of its wildness became synonymous with immorality (Baderoon 2017, 346). By contrast, "purebreds" like the German Shepherd, Rhodesian Ridgeback, and Boerboel came to represent the idea of racial purity (Doble 2020). In the postapartheid era, the inequalities between dogs, like their human owners, continue. "A dog is a dog. The average African dog is a little less than that," writes Chinodya (2017). "The average African dog is a creature to be kicked, scolded and have missiles thrown at it," while "a European dog is more than a dog …. It is a member of the family, with a personality, name, a kennel, a veterinary aid card and, of course, a budget." He continues: "A suburban African dog in an aspiring middle-class household is something between the two. While it probably benefits from the example of its white neighbors, it remains a household appendage." These sentiments were captured in 2012 when then-President Jacob Zuma chided upwardly mobile Blacks for loving their pets more than people, associating dog ownership with White culture (Hubbard and Brooks 2021). In the postapartheid era, dogs continue to play an important, albeit conflicted, role in urban society.

Insects are another frequent motif through which animal geographies overlap with South African urban studies. Byrne (2020) suggests that insects are used as metaphors for otherness, the *die swart gevaar* (the black danger), and related anxieties of White South Africans losing their Whiteness. They are wild animals that do not kowtow to the norms of modern society and social life. Parktown prawns are one such example of an urbanized insect. They sprouted up in the northern suburbs of Johannesburg in the late 1980s, drumming up fear in their human counterparts for three reasons: they are impervious to most insecticides, they have a habit of jumping at humans when scared, and they defecate when afraid. They were mostly eradicated from the suburbs by the hadeda ibises whose long beaks penetrated the ground and ate them (Byrne 2020). But the Parktown prawn remains a powerful metaphor – for example, prawns are fictionalized in Neil Blomkamp's 2009 film, *District 9*, in which the prawn is used as a metaphor for alien otherness. The film not only recalls the displacement of District Six in Cape Town by the apartheid government, starting in 1966 but also provides an example of eugenics through planning and control of urban space.

These examples from South Africa are by no means a complete assessment of animal mobilities. These discussions provide some reflections on the conceptual and empirical opportunities for thinking through the connections and disconnections between animal geographies and mobilities. It is also important to note that these categories are fluid with animal movements, policies, and imagery that intersects and overlaps across urban and African geographies.

Conclusion

This chapter highlights the opportunities for thinking about mobilities through the animal. It starts by outlining the theoretical rigor of animal geographies before taking an urban paws to reflect on how urban studies has been enriched by animal geographies. From there, this chapter calls for an approach to animal mobilities formed by bridging animal geographies and mobilities. Next this chapter turns to its focus on trans-paw-tation and the three tactics for reflecting on how animal geographies can contribute to mobilities – first by studying animal movements, second by considering animal legislation and policies, and third by examining the symbolism and metaphors of a combined understanding of animal geographies and mobilities. These examples present an approach to animal mobilities that can help expand current understandings of transportation failures in South Africa and beyond.

For transportation failure – and indeed geography as a whole – this chapter brings three factors into view: first, trans-paw-tation reminds us not to be ableist. Animal mobilities scholarship thinks beyond White, male, and able-bodied humans. It provides a voice to the other by featuring those who are typically excluded from the city. There is perhaps no discipline more White, male, and ableist than transport. And as a result of this limited approach, transportation success tends to be measured by a narrow populace – the expansion of electric scooter programs, for example, overlooks the needs of the elderly and car-sharing systems fail children and families. Just as animal geographies help distort the dichotomies between human and non-human, artificial and natural, domesticated and wild, trans-paw-tation repositions mobilities within a conceptual and practical otherness.

Second, trans-paw-tation provides an opportunity to think beyond normative metrics of success and failure, profit and debt, viability and sustainability. Animal mobilities illustrates a lens beyond capitalism and colonialism. Instead, it analyzes movement and stillness as regular and rhythmic parts of urban life. An approach to transportation that draws from animal geographies will think beyond traditional models of viability and service delivery, instead noting the myriad of ways in which transportation systems benefit users. A bus system designed and operated using a trans-paw-tation approach will not rely only on ridership and patronage to measure its success but instead see its operationalization as a means of transportation justice that benefits the entire city.

This leads to the final reflection – trans-paw-tation allows for broader thinking of what is possible. It exposes the rigidness of traditional transportation planning and its lack of creativity. When trying to plan abelist and viable systems, transportation geographers are incredibly limited. This broader approach provides boundless opportunities to imagine that anything is possible, encouraging the development and promotion of alternative models. This in return can benefit geography as a discipline by supporting thinking that reaches beyond the silos that we so often build and reinforce. Indeed, a trans-paw-tation approach imagines boundless opportunities to build a better city for all.

It is important to realize however that animal mobilities are not always productive. Indeed the spread of zoonotic diseases such as COVID-19 – the topic of Chapter 8 – as well as foot and mouth in the US and bird flu in the UK highlight the limitations of animal–human intersections and movement. For each of these illnesses, animal mobilities shift considerably – animal movement is loosened or restricted, policies on movement adjusted and symbolism reallocated. During the height of the Covid-19 pandemic, penguins waddled through Cape Town, kudu grazed in Tshwane, and the lions of Kruger Park relaxed on the golf courses and resorts. While human movement was limited, animal movement was enhanced. The pandemic drew attention to the ways in which animal mobilities are enabled and constrained by everyday life. An animal-focused approach to mobilities involves further reflection on animal–human and animal–animal connectivities and entanglements. A better understanding of animal mobilities henceforth promises to benefit both humans and animals, highlighting the failure of current systems to take account of living bodies whose ability to communicate is not always recognized. This study of transportation failures now applies these logics to transp-art.

7 Transp-art-ation

Introduction

A mural on the Sao Paolo subway station depicting the city's history, a sculpture at the Dubai bus terminal representing travel and movement, and a stained glass window in the Beijing train station featuring cultural imagery are all forms of transp-art-ation. The inclusion of art and aesthetics on transportation as metrics of success stands in contrast to mainstream measures of concrete and asphalt or fluidity and flows. Transp-art-ation makes art more accessible while simultaneously enhancing the visual appeal and cultural experience of mobilities. At the same time, it is costly and controversial (Amundsen 1995; Enright 2023; Hubbard, Smalley, and Hubbard 2010). Indeed, in South African cities, where transportation has historically been used to divide and control, art, in general, and transp-art-ation in particular, provides extensive opportunities for social and spatial transformation. This chapter considers the complex and controversial relationship between art and transportation in South African cities. In so doing, it considers how art geographies can contribute toward our ongoing discussion of transportation failures.

This analysis of transp-art-ation in South Africa contributes to the "ocularcentrism of geography" (Rose 2003), "the pictorial turn" (Mitchell 2006), and the agency of images. Geography is after all a visual discipline – illustrations, maps, photographs, PowerPoint, and videos all engage graphically with their audience. For Rose (2001), art has three sites of visualization – the production of the image, the image itself, and the "audiencing" of the image. Each suggests an alternative set of representations of who is engaging with the art and how they are doing so. This chapter applies Rose's (2001) "critical visual methodology" to transp-art-ation by considering the cultural significance, social practices, and power relations of the art, as well as the economic and political fabric of the surroundings and the people passing by. This chapter responds to geography's under-theorization of art by emphasizing not only the aesthetics of transportation but also the sounds, tastes, and smells through which art and transportation intersect.

By bringing analyses of art and transportation into geography, this chapter makes three critical contributions: first, it exposes the "material politics of images" (Bissell and Fuller 2017) in the global south. For African cities in particular, this chapter engages with "new critical pedagogies – pedagogies of writing, talking,

DOI: 10.4324/9780429322280-7

seeing, walking, telling, hearing, drawing and making – each of which pairs the subject and object in novel ways to enliven the relationship between them and to better express life in motion" (Mbembe and Nuttall 2004, 352). It builds upon ongoing discussions of art and aesthetics in mobilities and African cities; and the ways through which formal and informal public transportation systems utilize imagery, poetry, and sound to negotiate the economic, social, and political power (Agbiboa 2023; Hansen 2006).

Second, this analysis deterritorializes geography. Transp-art-ation demonstrates the inherent mobilities of art. Transp-art-ation is never static, always in motion, and always shifting in accordance with the user experience. And for art geographies, it adds a layer of complexity to discussion of public art, which is generally defined as murals, monuments, and sculptures, located in public spaces rather than museums and galleries; however, as urban space is oftentimes privatized and art and cultural spaces are to varying degrees public, this understanding needs to be expanded. For cultural geographers, therefore, the publicness of art is more social than physical, with scholars drawing examples from the experiential elements and the co-production of art (Guinard and Molina 2018). Public art equals public good: it beautifies the city and drives economic and social production. Critical art geographies scholarship has also studied the representation and symbolism of public art as well as the audience creating and critiquing it (Rose 2001, 2003). Transp-art-ation provides a further opportunity to reveal how the geographies of art are oftentimes mobile and dynamic.

Third, art geographies underscore the informality inherent to geography. This chapter draws attention to the gestures and movements of transp-art-ation, not only the murals and sculptures. And as such, it is attuned not only to the government-sanctioned programs and corporate slogans but to the everyday behaviors and languages through which everyday South Africans move. Indeed, it draws on discussions of street art, for instance, which uses the surfaces of buses, trains, and subway cars as canvases for creative expression. In the 1980s, the New York City Subway became a frequent site for graffiti. This subversion of traditional artistic spaces not only challenged societal norms but also prompted a reevaluation of the role of public transportation in urban life (Fitzpatrick 2009). Murray (2014) explores the relationship between art and mobility in Belfast, a city long dominated by conflict, by illustrating how murals make sectarian rivalries visible. He goes on to suggest the peace walls and painted curbs are not only material artifacts but also representative of everyday experiences of mobilities. In so doing, this chapter demonstrates the many intersections between movement and stoppage, inside and outside, inclusion and exclusion. Indeed, transp-art-ation is deeply embedded in the form and function of the city. This discussion of transp-art-ation brings additional evidence of the possibilities of, and need for, further academic overlapping.

The next section focuses on public art in South Africa and outlines its central role in postapartheid reconciliation. From there, this chapter unpacks the overlaps and intersections between art and transportation geographies. It provides a conceptual framework through which to examine transp-art-ation in South Africa. Next, this chapter develops a set of critical approaches to transp-art-ation: art on transportation – evident through the art infrastructure built in bus and train stations, mobile art – evident through the art adorning the vehicles even as they move, and representations of transportation – evident through abstractions that narrate our movement. In conclusion, this chapter reflects on the importance of bringing art geographies into conversation with transportation geographies.

Art in Postapartheid South Africa

South African cities have long held a complex relationship with art: San rock art featured drawings of humans walking and running while 19th-century colonial South African cities could scantly be described as artistic or cultural. The *Uitlanders* were crude colonists and their strict religious views believed art to be the work of the devil (Shorten 1970). Later, under apartheid, art was used to solidify White power, while Black bodies and their aesthetic were systematically excluded from the city (Bunn 2008). The end of apartheid, in contrast, was accompanied by a frenzy for art. Some of the art was commissioned by the government, some by private institutions, and some emerged unplanned (Harrison 2014). Public art was typically created by Black artists, depicting African beliefs and memories as well as contemporary culture. These memorials, monuments, and murals became a visual strategy through which to bring Black aesthetics and Black stories back into the urban narrative. Meanwhile, historical statutes – for example, Pretoria's statue of Paul Kruger or Cape Town's statue of Cecil Rhodes – typically remained standing with new pieces added alongside. This stitching together of past, present, and future is extremely controversial and evident across the art world.

Much of the postapartheid public art was explicitly political: in Durban for example, in 1992, a multiracial team of artists in Durban painted the Human Rights Wall, depicting the clauses of the Universal Declaration of Human Rights on the perimeter wall of the former Central Prison where antiapartheid activists had been held. And, in 1994, the Nomkhubulwana mural in Durban was painted days before the election depicting the informal traders typically present in the market. This mural was used to legitimize the presence of the informal traders and highlight their role in the economic and social life of the city (Marschall 2008).

Other art pieces were economic: in 2004, for example, Cell C, then South Africa's newest mobile phone operator, sponsored 30 murals across Johannesburg. The company instructed each of the artists to incorporate the phrase "…for the City" and include some C-shaped forms – a half moon, birds flying in a C formation – in the mural (Marschall 2008). This project is particularly controversial as the goal of the project was ultimately commercial.

Public art thus became a heavily contested neoliberal tool for economic development and urban regeneration (Guinard and Margier 2018). Policymakers saw art as a strategy through which to address urban decay, extreme inequalities, and a declining economic base. And public art was ideal as it was also associated with postapartheid reconciliation and social development. In Johannesburg, for example, slogans like "A World Class African City" were part of a citywide strategy to merge economic growth with social development; and arts-based events such as the Johannesburg Art Fair and the 2010 FIFA World Cup, and cultural spaces like the Apartheid Museum, the Newtown Precinct, and the Soweto Theatre, to name just a few, were expected to fulfill redistribution goals (Guinard 2018).

Many of these initiatives were city-led and city-financed. Both Cape Town and Johannesburg have a mandate to spend 1 percent of the construction budget of all major city building projects on public art (Department of Community Development 2006; Espi 2017). The 2006 Public Art Policy in Johannesburg, which dedicates 1 percent of all building projects over R10 million for art, claims to provide Johannesburg residents with "shared symbols which build social cohesion, contribute to civic pride and help forge a positive identity for the city" (Department of Community Development 2006, 1). Policymakers justify this expenditure by suggesting that public art supports the creative industries that, in turn, create work for artists, designers, and fabricators. On a larger scale, public art catalyzes economic growth by attracting investment and tourism. Between 2006 and 2012, the City of Johannesburg spent nearly R20 million commissioning 680 pieces of public art including the *Eland* in Braamfontein (van den Berg 2007), the *Angel of the North* in Hillbrow (Luthuli 2010), and the *Firewalker* on the Queen Elizabeth Bridge (Kentridge and Marx 2009) (Harrison 2014).

Entire neighborhoods were redeveloped during this postapartheid frenzy for public art. Johannesburg's Newtown is one of the largest city-led arts-based urban redevelopment projects. The Market, an informal cultural quarter developed in the 1970s by artists alongside their anti-apartheid activities, was rebranded Newtown Cultural Precinct by the City's Arts, Culture and Heritage Department in 1991 and the Johannesburg Development Agency (JDA) in 1998. The JDA turned Newtown into the cornerstone of a city-led approach to cultural consumption and commercialization. In the early 2000s, the JDA commissioned 560 carved wooden heads installed on bollards as part of a broader strategy to launch a cultural district in the area (Ho 2018). Despite these attempts at arts-led regeneration, Newtown as a creative district never quite happened. Minty (2006, 2023) criticizes this approach, concluding that art districts like Newtown drive capital improvements by targeting international investment and tourism instead of local community and social development. Minty also argues that Newtown in particular failed when it turned from a site of cultural production to a site of cultural consumption.

In addition to political and economic reasons, public art was commissioned to make "symbolic reparations" and commemorate South Africa's painful past (Minty 2006). The *Eland* (2007), a five-meter concrete sculpture designed by Clive van den Berg, which stands on the corner of Jan Smuts and Ameshoff Street in Braamfontein, represents the traditional relationship between animals and the land, while the *Angel of the North* (2010) by Winston Luthuli, which stands at the corner of Kotze and Queens Street in Hillbrow, holds his arms outstretched like Christ the Redeemer. Perhaps, none are as symbolic as The *Firewalker*, a ten-meter steel silhouette of a woman carrying a brazier on her head. It was designed by William Kentridge and Gerhard Marx and commissioned by the JDA in 2009. The site, on the corner of Simmonds and Bicard Street, on a traffic island at the foot of the Queen Elizabeth Bridge, holds both symbolic and economic power. It is located on the former site of an informal trading market and taxi parking. It used to be common to see women cooking and selling food to busy commuters until the city "cleansed" the area in anticipation of the 2010 FIFA World Cup. The *Firewalker* thus stands as a memorial to the informal city (Emmett 2021).

In some instances, public art became associated with sociopolitical resistance. The downing of a statue of the British imperialist Cecil Rhodes on the University of Cape Town campus in April 2015 reveals the power of art to fuel broader debates. In March, the University of Cape Town student Chumani Maxwele held a protest at the statue's base and threw a portable canister of human feces at the statue. The protest, later dubbed #RhodesMustFall (Elliot-Cooper 2017) then spiraled into the national and international movement #FeesMustFall (Habib 2019). The protests, which began by opposing the high costs of university education, later turned into opposition to the enduring legacy of colonialism, racism, and inequality in wider society. The statue of Rhodes became "a literal canvas conveying the various sentiments of students for and against its removal," explains Gurney (2018, 35). After its downing, the statue was wrapped in black plastic garbage bags and bound with tape and the plinth turned into a rotating art installation, first with an empty plastic chair to symbolize the former role of Rhodes in particular and White power more generally. Gurney (2018, 33) calls the afterlife of the statue's plinth "a zombie monument" concluding that "visual art has the capacity to recalibrate the public sphere by acting as a vector of both resistance and radical re-imagination." In this instance, art exposed the deep ruptures in society and invited dialogue for positive change.

Not all public art, however, has fueled economic, political, or social change – and in fact, much of it has failed. Indeed, many South Africans were left out of this art-led transformation – especially artists working beyond the mainstream studios and those living in the underdeveloped areas of the townships whose peripheral location was beyond the focus of government-led redevelopment strategies as well as urbanites whose experiences were not featured in the typical postapartheid shifts (Perryer, Koloane, and Mosaka 2004). Guinard (2018, 118) criticizes the postapartheid boost to public art efforts, claiming that installations were generally "permanent, installed in big cities, produced as part of top-down strategies and made for an abstract and imagined public rather than with or by actual and existing publics." Likewise, many of these artworks have been poorly maintained. The *Firewalker* is just one example of many public artworks in Johannesburg that have become neglected and in their current state used as a urinal (Emmett 2021).

The rationale supporting the postapartheid frenzy for public art is complicated. In many instances, art in South Africa has been used in diametrically opposing ways – as a means for redistributive justice that still focuses on the economically advantaged, as a tool for urban redevelopment that also leads to displacement, and as a driver of social reconciliation that is also symbolizing colonial power and anti-establishment resistance. Further reflection on these contrasting forces suggests the importance of bringing art into discussions of transportation. Indeed this focus on transp-art-ation reveals the procreant nature of failure by uncovering the politics of images/ry, de/territoriality of art, and the intrinsic informalities that constitute the city.

Bringing Art into Transportation

Art and transportation geographies, while seemingly distinct, share intricate and overlapping historiographies. Art and movement have long interrelated: there are paintings of chariots and ships on Egyptian hieroglyphs, and Greek and Roman pottery was adorned with horse-drawn carriages; during European industrialization, 19th-century paintings depicted steam engines billowing with smoke, alluding to the majestic presence of trains; and the impressionists captured the shifting mobilities brought about by the automobile. In literature, Jules Verne and H.G. Wells imagined submarines, rockets, and flying machines before these were invented. In developing a framework for understanding transp-art-ation, this discussion delves into the multifaceted connections between art and transportation, highlighting how art has been inspired by various forms of mobility, and in turn, how mobilities have been shaped by art.

In the early 20th century, governments and private companies sponsored and financed art on transportation infrastructure. The London Underground (Bownes, Green, and Mullins 2012), New York City Subway (Fitzpatrick 2009), and Boston trolleys (Breitbart and Worden 1994) each included attention toward art and aesthetics in the original design. Boston's Massachusetts Bay Transportation Authority was, however, the first American city to add art to its subway stations when it launched "Arts on the Line" in 1978 (Skarjune 1989). And in New York City, Bill Brand's *Masstransiscope* (1980), which includes 300-foot-long paintings that link together and form an animated film as passengers ride through DeKalb Avenue on the Q or B Manhattan bound train, is one of the earliest examples of mobilized transp-art-ation. At this same time, however, transportation in mid-20th century became more utilitarian emphasizing efficiencies over aesthetics (Hubbard, Smalley, and Hubbard 2010). And, art on transportation became less conventional and less legal – buses and trains became canvases for self-expression and social commentary. These forms of graffiti linger today in informal transportation systems where art represents both the economic and social identity of the vehicle and its driver.

Nowadays art regularly appears in airports, bus stops, subway stations, highway rest stops, and traffic circles. It ranges from the trifling to the iconic, encompassing statues and murals as well as both legal (e.g., signage and marketing) and illegal (e.g., graffiti) forms of communication. Transp-art-ation is thought to enhance the aesthetics of the system thereby boosting ridership, increasing passenger comfort and safety, helping to integrate the system into the neighborhood, and encouraging support for public transportation in general. These benefits extend from the system into the neighborhood and throughout the city (Amundsen 1995). For Enright (2023, 71), art serves as a panacea for "sometimes-complementary and sometimes-conflicting policy objectives" ranging from public safety to neighborhood revitalization, business and economic development, social reconciliation, and environmental objectives. Enright (2023) subsequently questions the extent to which transp-art-ation can reach all of these objectives.

Transp-art-ation is funded through both state and local art programs in the form of departmental appropriations, tax increment financing, and hotel and sales tax as well as private sponsors and grants. Many governments around the world maintain a 1 percent policy for public art programs: for example, Boston, Chicago, Dallas, Los Angeles, Montreal, New York, Philadelphia, Phoenix, Portland, Seattle, and St. Louis in the United States alone, as well as analogous programs in Europe (Barcelona, Brussels, Lisbon, Munich, Naples and Warsaw), in Latin America (Buenos Aires, Mexico City, Santiago and Sao Paolo), in Asia (Beijing, Delhi, Dubai and Singapore), and in Africa (Cape Town and Johannesburg) (Enright 2023). This financial approach began in the 1930s through US government-led workfare programs like the New Deal that dedicated 1 percent of the costs of any publicly funded building or infrastructure project to acquire, commission or restore works of art (e.g., paintings and sculpture). Although the program ended in 1943 due to the ongoing World War, interest continued, and in 1959, the City of Philadelphia was the first postwar city to fund a percent-for-art program. Transp-art-ation programs have since spread around the world (Hubbard, Smalley, and Hubbard 2010).

Within these discussions of transp-art-ation, a geography lens creates three additional layers: the importance of representation, the space in which it sits, and its audience. For Anderson (2019) and other cultural geographers (Hall 2007; Rose 2001), representations matter. Anderson (2019) recognizes the "force of representations" and their "capacities to affect and effect, to make a difference" (2019, 1120). He cautions, however, that this power is both uneven and unequal. "Artful cities" (Hall 2007), then, are those that draw their economic, political, and social character from art. This is especially apparent in the recent neoliberal context in which cities are being made and remade through a commercial lens. Anthony Gormley's *Angel of the North* (1998) in Gateshead, for example, was promoted by the local authority as an economic redevelopment tool with symbolic value, but its high costs in an otherwise disenfranchised region in postindustrial England make it controversial. Indeed, representations matter in cities – but how they matter is often unclear and contested.

The relationship between art and space is also important for geography – what Hawkins (2013, 57) calls "art-site relations" to describe "tensions between the sites of art's production and consumption and those it represents or evokes." This spatializing of art geographies, which examines the relationship between art and the people making and remaking place through art, was as evident in impressionist paintings as in today's art exhibitions. Today art has become de-spatialized with a less fixed, more fluid physical locatedness. Public art happens beyond the topographical and topological inside – the gallery, the museum – and can be seen, experienced, and critiqued across the city. And this de-spatializing of everyday creativity is also ephemeral, extending to include festivals, parades, and other expressive arts. The artist then becomes both anthropologist and ethnographer and their practice becomes an intervention into urban life.

Geographers, however, are not only concerned with understanding the meanings of images or the space they inhabit but also how they intersect with their audience. Rose for one advocates for a "critical visual methodology" that considers the "visual in terms of the cultural significance, social practices and power relations in which it is embedded" (2001, xv). She highlights the role of the audience by noting their interactions with the image, how the audience makes art meaningful, and the wider context into which the two are embedded. For Bissell and Fuller (2017), representations are an "image-object" through which mobility infrastructure is enacted. In their analysis of a private highway in Sydney, they consider not only the material agency of images but also the audience. Instead of interpreting maps, visualizations, and other renderings only as "developer-propagated mobility ideologies" (2017, 2478) however, Bissell and Fuller bring the audience back in, arguing that mobile bodies are also part of the machinations that shape mobilities.

The critique of transp-art-ation in the following section adds to these discussions by considering how transportation failures are made and remade through art and in return how art reveals the productive nature of failure. It discusses art on transportation in its most literal sense as well as how art serves as a language of mobilities, a coping mechanism for long commutes and irregular services, and how it can brighten an otherwise dreary ride. This analysis of transp-art-ation in South Africa invites further reflection in three specific ways. First, transp-art-ation highlights the broader intersections between art, transportation, and urban geographies. Rather than honing in on the design, engineering, or operations of transportation systems, art draws attention to the cultural, political, and social characteristics of both the system and the city – what Bissell and Fuller (2017) call the "material politics of images." These politics not only feature movement but also reflect the barriers to mobility, where stasis is warranted, and how to negotiate alternative approaches to mobility. Transp-art-ation itself is always political, never static, and therefore what appears as a failure can still be both dynamic and fertile.

Second, transp-art-ation moves beyond the fixities and flows of transportation. It can be both territorialized in a particular station and mobilized on the vehicle. Transp-art-ation needs to be explored not only through the murals and sculpture at bus stations but also on the vehicles moving and the ways that the public performs and inhabits that mobility. In so doing, transp-art-ation showcases the mobile nature of urban life. It also recognizes the voices of the public in transportation. Whereas transportation engineering typically neglects the role of users in system design and management, transp-art-ation exposes the central role played by users both on and off the system. A transp-art-ation lens on failure provides a means by which to reflect on the publicness of transportation, and its benefit to the local community as well as the city. It underscores the role of the people riding the buses, hailing the taxis, and waiting in traffic. It features the person, their behaviors, conjectures, and communities, as central to transportation and urban life. Transp-art-ation thereby humanizes transportation and helps it engage with a broader audience.

Third, transp-art-ation exposes the informal characteristics mobilities. In formal transportation modes, transp-art-ation serves as a silent reminder of the informality inherent to the city. The very act of reserving a space for art on transportation can be seen as rebellious. By building, designing, financing, and maintaining murals and sculptures in otherwise austere, modernist settings, transp-art-ation decenters the role of the prescribed and proper aspects of transportation. And, by extending beyond the circumscribed structures of transportation – the train schedules, the bus stops – art embeds informality within the broader urban context. In informal systems, transp-art-ation is a central feature – the slogans and sounds enhance the economic and social character of transportation vehicles, attract customers, and provide drivers with a creative outlet. Riders oftentimes recognize their favorite vehicle from its exterior and wait for it; other times, music might play a role in this art. This explains why transp-art-ation features heavily across the informal sector. In formal and informal modes, transp-art-ation exposes the unauthorized and unstable aspects of urban life. However, the overlaps between transp-art-ation and informal transportation remain scant. In analyzing both the formal and informal characteristics of transp-art-ation, this analysis highlights how transp-art-ation can provide an alternative understanding of transportation failure – a discussion we subsequently turn to in the next section.

Transp-art-ation in South Africa

Transp-art-ation – a mural on the wall of a train station, a sculpture in a bus rapid transit (BRT) station, and hand gestures to communicate with a moving taxi are all forms of public art closely associated with transportation and mobilities (see Figure 7.1). Ongoing discussions of transp-art-ation typically focus only

Figure 7.1 Mural of Rea Vaya bus on Louis Botha Avenue in Johannesburg (photograph by author, November 2019). This artistic representation painted onto a building on Louis Botha Avenue along Johannesburg's Rea Vaya Phase 1c illustrates an example of transp-art-ation.

on the formal sector – buses and trains – without considering informal and non-motorized movements. Likewise, the research typically only considers fixed art, overlooking the movement inherent in transp-art-ation. Lastly, existing research on transp-art-ation is overly focused on the global north – thinking about formal transportation infrastructure and government-sponsored initiatives. But, in many cities of the global south, transportation is provided by informal or semiformal systems and therefore transp-art-ation takes an entirely different approach.

These examples from South Africa provide an opportunity to critically reflect on transp-art-ation and its role in the economic, political, and social life of the city. It brings examples that are not unique to South African or global south cities but are suggestive of more informal approaches to transp-art-ation. The first section, art on transportation, provides stories not only from art on South Africa's newly-built BRT stations but also from taxi ranks, which suggests a complex transportation landscape. The second section, on mobile art, discusses how transp-art-ation is set in motion, focusing on the informal sector and South Africa's minibus taxis. These examples not only highlight the central role the informal sector plays in economic development but also how it is leading transp-art-ation. The third section, on the representations of mobilities, discusses instances where transp-art-ation transcends the sector to become part of broader cultural practices.

Art on Transportation

The installation of public art in transportation stations in Cape Town, Johannesburg, and Nelson Mandela Bay represents the most mainstream approach to transp-art-ation. In South Africa however, transp-art-ation became integral to the introduction of BRT, serving as a tool by which to provide dignity to commuters and bring culture back to the city: in Johannesburg, art is a mainstay of the city's Rea Vaya BRT system; in Cape Town, art adorns all of the MyCiti stations; and in Tshwane, decorations sit atop the vertical columns at each A Re Yeng station.

Art was integral to the design and development of Johannesburg's Rea Vaya Phase 1A when it launched in August 2009 as South Africa's first BRT system (see Figure 7.2). Each of the 30 stations along the 25.5-kilometer route operating between Thokoza Park and Ellis Park includes an original commissioned etching. The artwork was sand-blasted onto the steel and glass station, with the intention of displaying art not only to Rea Vaya users but also to the general public. Lael Bethlehem, CEO of the JDA (2005–2010), explained: "Every station has an art program and that was part of the flagship of Rea Vaya" [47]. The art was designed to create a unique identity for each station while simultaneously strengthening coherence across the system. "This has been a great opportunity to bring original Johannesburg art to a wider audience," Steven Sack, Johannesburg's Director of Arts, Culture and Heritage (2004–2013), reasoned. "It has also created opportunities for many artists in the city." Rehana Moosajee, the Mayoral Member for Transport (2006–2013) responsible for implementing Phase 1A, explained: "The inclusion of these exciting artworks contributes to the vision of a public transport system that brings dignity and excitement to the everyday experiences of our passengers" (Arrive Alive 2009).

Figure 7.2 Glass etching on Rea Vaya station in Johannesburg (photograph by author, January 2020). Each of Johannesburg's Rea Vaya BRT stations features a glass etching depicting the area's distinct architecture, history or people.

The designs typically depict mobility and movement as integral to urban life. The portraits of commuters on Mine View Station, a tricycle depicted on Westgate Station, the busy streets painted on Carlton Station Eastbound – each portray life in Johannesburg as hectic and hurried. Some of the art is symbolic – the design of Old Synagogue Eastbound, for example, accounts for architectural similarities between houses of worship; other designs are functional – for instance, the design of Art Gallery Station uses a digital trace of the Google Earth map of the station's location. Several of the designs incorporate postapartheid approaches to reconciliation – "Not Just a Number" at Ellis Park East station, for example, depicts stick figures recalling the simplicity of children's drawings as well as the rock drawings of historical peoples. Stations also consider the audience – Chancellor House Station Eastbound, for example, includes bubble narratives which can be completed by commuters; and at Orlando Police Station, artist Sipho Gwala displayed closely cropped photographs of residents of Alexandra taken by visiting researchers from the University of Johannesburg. The design suggests that researchers should be more sensitive to their research subjects. All of these designs give voice to both transportation riders and the general public.

Likewise, in Cape Town, 50 artworks adorn all 42 stations of MyCiti Phase 1, which opened in 2011 (see Figure 7.3). "Showcasing local art is synonymous with excellence in public transport the world over, and for many people their only exposure to formal art is through its provision in public spaces and on public infrastructure," writes Brett Herron, Mayoral Committee Member for Transport and Urban Development (2011–2018) in the Foreword to the City's transp-art-ation publication (Espi 2017, 3). He reasons that art

> contributes to our society by giving expression to our history, our dreams, hopes and aspirations in the most creative of ways. And it contributes to our economy when we support artists to make a livelihood based on their unique talents and abilities.
>
> (Espi 2017, 3)

Like in Johannesburg, Cape Town dedicated 1 percent of the budget of MyCiti for transp-art-ation.

Transp-art-ation on MyCiTi comes in many forms: mosaic at Gardens and Stadium stations; text at Airport and Century City stations; painting at Section, Vrystaat,

Figure 7.3 Art on MyCiTi station wall in Cape Town (photograph by author, January 2020). In Cape Town's main MyCiTi interchange, as in all the stations in the system, artwork beautifies the transportation experience.

and Porterfield stations; murals at Atlantis, Omuramba, Mitchells Plain, and Gardens Stations; sculpture at Civic Centre, Wood, and Grey stations; illustration and graphics at Potsdam, Circle East, Century City, Civic Centre, and Thibault stations; infographics at Milnerton, Racecourse Lagoon Beach, and Parrden Eiland stations; and photographs and photomontage at Neptune, Sandown, and Killarney stations. Many of the stations depict means of transportation (e.g., the art on the Civic Centre Station portrays the uitspan and the bicycle), people in transit (e.g., the figures on Vrystaat Station), animals on the move (e.g., the herons at Atlantis Station), and the importance of the station as a meeting place (e.g., a painting at Section Station of people coming and going and encountering each other at the station). Some of the stations depict the atrocities of apartheid (e.g., the forced removals on the Woodstock Station), but many focus on postapartheid restitution and reconciliation. Some artists directly include the viewer in the design – Airport Station, for example, presents little scraps of history with quotes from across time – while others consider the exterior as well as the interior view to cater to pedestrian-viewers and others experiencing the art from the outside. The art on MyCiti was designed to connect commuters with the stations, to generate interest in the economic and social happenings around the stations,

Figure 7.4 Art at the Bree Street Taxi Rank in Johannesburg (photograph by author, March 2014). In 2019, the Bree Street Taxi Rank, one of Johannesburg's busiest transportation interchanges, was beautified by Kgomotso Neto's photographs of his daily commute.

and to enrich commuter experiences with cultural infrastructure. Herron explains: "The art portrays the cultural heritage of Cape Town and gives commuters a sense of the history of the neighborhoods they are travelling through and the culture and roots of those living there" (Espi 2017, 3).

Transp-art-ation is not only installed in formal transportation systems, however; in South Africa, it is also installed in minibus taxi ranks where thousands of minibus taxis pick up and drop off passengers daily. In 2019, photographer Kgomotso Neto hung three of his images from a series entitled *Middle Centre* on the walls of the Bree Street taxi rank in downtown Johannesburg (see Figure 7.4). These photographs, which illustrate Neto's daily commute, were viewed free of charge, in the space in which they were snapped. "Art is usually hung on white walls in fancy galleries," explains Neto, "but I think it's quite important that we make this work accessible to everyone" (Maputla 2019). This approach to transp-art-ation brings art to the commuter, providing a cultural break in the midst of an otherwise informal hustle and bustle. It is also suggestive of a broader approach to transp-art-ation found especially in South Africa.

Mobile Art

Mobile art is an approach through which transp-art-ation is set in motion. In South Africa, this occurs through projects that apply art to public transportation vehicles. These art installations showcase multimedia, photography, performance art, poetry, video, and visual art on the moving vehicle. They tackle current and cutting issues of race and identity, working conditions and inequalities, and other matters relevant to postapartheid South Africa. Most importantly, they make art accessible beyond museums and galleries while simultaneously making transportation more sociable.

Minibus taxis have long served as mobile art, depicting popular culture icons and events. In the 1970s, in opposition to the banning of American comic books, taxi drivers decorated their taxis with popular superheroes like Mighty Man and Luke Cage (Woolf 2013). In the 1980s, one of the earliest taxis was nicknamed the "6 Mabona" which translates as "six eyes," to characterize the six big headlights of the Chrysler. The Toyota Super 10 was renamed the "Zola Budd" after the South African runner Zola Budd Pieterse, while the Isuzu was renamed the "Mary Dekker" after the 1984 Olympics runner who fell after a collision with Zola Budd, and more recently, the "Semenya" was named for Caster Semenya, the Olympic gold medalist runner (Classen and Van der Post 2020). Drivers also proudly display the logos of local football teams such as the Orlando Pirates and the Kaizer Chiefs.

In the past decade, however, the minibus taxi industry has been energetically driving mobile art: in 2014, SA Taxi, a company that finances the minibus taxi industry, launched the SA Taxi Foundation Art Award, to provide emerging artists with opportunities to showcase their work. Each year, the work of the six winning artists is displayed for six months on ten taxis as they operate around South Africa, especially in arts districts like Maboneng, Johannesburg (SA Taxi 2017). The SA Taxi Foundation art award aims to provide opportunities to artists while also ostensibly attracting riders and future advertising revenue. Queeneth Brown, coordinator of SA Taxi's social investment strategy, explains, "the taxis running on conventional routes will promote the finalists' work to people who would not otherwise have the time or funds to go to a creative hub or to an art gallery" (Hlalethwa 2018).

Also in 2014, as part of the Joburg Photo Umbrella, a festival sponsored by the Market Photo Workshop, six Johannesburg Bree Street taxis operating between Dobsonville and Soweto were wrapped in Jerry Gaegane's *Marang a Letsatsi*, a series of photos documenting the living conditions of informal miners (Hlalethwa 2018). This mobile art installation illustrates an engagement between the community photographed, the community of riders, and the community of witnesses. Rather than detaching the communities, Gaegane's mobile art integrates them.

In Cape Town, as part of the public arts festival *(Un)Infecting the City* presented by the University of Cape Town Institute for Creative Arts, Thania Petersen created what she calls "an immersive art experience" (Bantom 2021). Petersen added a 22-inch screen and sound system to a Quantum taxi that traverses the route between Cape Town city center and Hanover Park. Her film, *Kassaram*, which discusses the complexities of Cape Town's Cape Malay and Coloured identities, was then played inside the taxi, which was also covered in art from the film (Tyilo 2021). For Petersen, "the idea is to make art more accessible to people in disadvantaged communities free of charge." She explains, "South Africa has many extraordinary artists, but most of their work is in private collections and museums." People from disadvantaged backgrounds therefore "don't have access to art because they cannot afford it" (Bantom 2021).

> The commuter will be immersed in art. For me [Petersen], our public transport system reveals so much of the inequality we face daily. It exposes how our social systems fail us with reliable infrastructure, we can't depend on decent public education, hospitalisation, trains or buses. I think I could safely estimate that 99 percent of our public transport commuters in this city is of colour from the areas apartheid had dumped them into. They can not afford to take a day off to visit galleries or the theatre or pay to get into a private museum. So what I am hoping is to take it to them and to enrich their lives with art at any given time of the day.
>
> (Naidoo 2021)

The taxi owners "did not think twice about getting involved, because they saw it as an opportunity for art and transport to meet" (Bantom 2021).

These examples of mobile art demonstrate the possibilities for art and transportation geographies to extend art to the underprivileged while simultaneously providing marketing for the taxis. It enables busy commuters to experience art. They therefore also reveal the possibilities for transp-art-ation to serve South Africa's pro-poor agenda.

Representations of Transportation

Transp-art-ation also includes abstractions, expressions, and other representations of transportation. This art is oftentimes informal, not codified or formalized, but it provides a language that supersedes art on transportation or mobile art. This approach not only reveals how the city is shaped by transportation but also how transp-art-ation is shaped by the community of users and their daily performances. It also exposes the intersections between formal and informal, stasis and movement, as well as broader economic and social conditions inherent to urban life.

Photographs, for example, provide a powerful tool through which to represent transportation. In 1967, Ernest Cole captured the Black commuting experience, complete with overflowing train stations and overcrowded trains. His photographs were the first to portray "washing" – the practice by which passengers hang off a fast-moving train, resembling clothes on a clothesline (Cole 1967). More recently, in 2020, Dan Carter's photographs of modern-day pantsula taxi drivers and owners are another representation of transportation. In the series, *Modern Day Taxi Owner – An Ode to the South African Taxi Industry*, taxi drivers and owners are photographed dressed head to toe in the latest fashion and sitting in their taxis inviting customers to take a ride. These photographs exhibit "the gentlemen of the taxi industry" in contrast to their usual portrayal as rude and unhygienic (Zakwe 2020).

Hand signals, a gestural language by which commuters signal to their minibus taxi, are another representation of transportation and mobilities. Between 2004 and 2007, Susan Woolf documented the taxi hand signals she observed on Johannesburg's streets. Some hand gestures are literal – for example, two pointed index fingers crossed to represent crossroads or a curved thumb and finger carving out the letter C for Cresta – while others are symbolic, for example, cupping an imaginary orange to signal to taxis driving to Orange Farm or the performance of one hand dipping downward and then upward to signify the bumpy roads in Diepsloot (Woolf 2013). In 2007, Woolf presented these signals in a series of 26 gouaches of gloved hands making each sign, and that same year, she also published her taxi hand signals as a book (Woolf 2007). In total, Woolf has identified 50 hand signals, in what Molefe Tsele describes as South Africa's 12th official language, "the language of commuters" (Woolf 2007). However, there is nothing formal about this language – instead, it was developed organically by users. Allara (2017) suggests that these hand signals are "evidence of the resilience and resourcefulness of South African commuters." It represents community and communication. It is also suggestive of opposition to White representations of transp-art-ation – for example, the hand signal for KwaThema, meaning "rest in peace," is performed when a commuter holds their hands together resting on the left side of their face. In some Western cultures, this signifies going to sleep but in South Africa, the taxi sign does not match the precise meaning of going to sleep but rather a linguistic adaptation expressing either their lack of rest due to violence or the frequent deaths that result from the violence and their longing for rest.

Abstractions also include movements such as dance, gestures, and pantomime to symbolize mobilities and their experiences. In 2017, Nadeem Diempie Swartz, a taxi guard in Cape Town, became a YouTube sensation when videos of him dancing in and out of his taxi went viral. Taxi riders and other road users were thrilled to see him gyrating, prancing, and twirling as the taxi slowly navigated traffic, honking and applauding his efforts to make the drive more enjoyable. Describing his motivation to dance, Swartz said, "I just decided to ask the driver to turn it up; I opened the door and started to dance. I didn't know these moves would make me famous" (Petersen 2017). This form of transp-art-ation is ephemeral, occurring in one moment and then disappearing in the next, leaving only the viewer with an enduring memory.

Other times, expressive arts are codified in video and visual arts. In 2014, the *Taxi Dance* was popularized by the Boyz N Bucks collective and was later seen in Cassper Nyovest's music video *Gusheshe* (2013). In the video, the dancer mimics the driving habits – the beeping, the careless turning of the wheel, the slouched backs in their seats, and the quick jerks after taking off at top speeds – that are typical in South African taxis. The dance was then immortalized on the Cartwright taxi rank in central Durban by Wesley van Eeden (a.k.a., Resoborg). As part of a revitalization of the taxi rank in 2017, the eThekwini Metropolitan Municipality commissioned Resoborg to paint a mural. The mural, on Umgeni Road, was meant to depict the intersections between drivers and riders, to represent the community that develops as the two, the transporters and the transportees, share space (Hlalethwa 2018).

Together these, three approaches represent a more inclusive approach to transportation – they feature movement and stillness, fleeting and permanence, and insider and outsider experiences. This arts-based lens extends our understanding of transportation not only as a system by which people move about the city but as a form of cultural expression with tremendous symbolic value that extends mobilities beyond the confines of the system and the city. Indeed, this broader lens rejects normative measures of success/failure instead trying to contemplate the experiences of mobilities and how it reflects urban life.

Conclusion

Much has changed since the urbanist Jane Jacobs (1961) famously declared that "a city cannot be a work of art." Today, cities are teeming with public art. Murals, monuments, sculptures, and gardens ostensibly generate social goods ranging from job creation and economic growth to civic pride and social cohesion (Hawkins 2019; Rose 2001). In many instances, public art serves as a tool for culture-led regeneration, while critics of public art policies contend that art has been commodified and is associated with gentrification and displacement. These controversies are perhaps most apparent in transportation where buses and bus stations are now routinely draped in murals and sculptures that policymakers claim enhance the commuting experience and catalyze urban growth. It is therefore fitting to think about how and where art geographies and transportation geographies overlap and intersect within urban geography.

In conclusion, this chapter has argued that bringing art into transportation geography exposes failure as procreative. Across the three approaches discussed in this chapter – art on transportation, mobile art, and representations of mobilities – transp-art-ation highlights the urbanness of transportation, informality within the city, and the role of urbanites in the city. It brings discussions of representations to the global south, deterritorializes transportation geography, and underscores the informality in transportation to geography. In particular, this chapter reveals how visualizations can expose alternative framings of the city. And, in so doing, it brings additional evidence of the ways in which transportation failures can be reimagined and revalued. The next chapter applies these approaches to a study of what happens when transportation fails.

8 When Transportation Fails

Introduction

This chapter considers what happens when transportation fails. It explores the COVID-19 (hereafter Covid) pandemic in South Africa and the subsequent mobility restrictions introduced to stymie it. By drawing on a longer history of disease in cities, it underscores the critical connections between health and mobilities especially as it relates to virus transmission. In doing so, this chapter notes that transportation failures are also global failures.

In December 2021, South Africa once again became the epicenter of the Covid pandemic. A new variant, dubbed Omicron by the World Health Organization (WHO), had been identified. Within days, the world shut its doors to South Africans, canceling flights and visas for international travel and turning South Africans into global pariahs. Internal travel was similarly restricted and movement was locked down. This episode highlights the ways in which mobilities and transportation are inherent to our lives and what happens when those mobilities are cutoff. This chapter engages with the long history of disease and the city to make sense of the Covid pandemic. In particular, it investigates how the pandemic has impacted mobilities and vice versa. It explores the overlaps between mobilities and Covid by reflecting on restrictions on all forms of movement, including international mobilities, and urban mobilities.

In so doing, this chapter makes two important contributions to the burgeoning literature on Covid and transportation. First, it locates the discussions in South Africa, a country that has been at the forefront of the pandemic: it is one of only a few countries to sequence two variants – Beta and Omicron; only 65 percent of the country is vaccinated; and to date, there have been more than four million cases and the official death toll is over 100,000 (Worldometer 2024a). This chapter reflects on the steps taken to prevent the spread of Covid. It focuses on South African cities where the pandemic has hit the hardest. In so doing, it contributes to the scholarship that hopes to find lessons from South Africa's past experiences with pandemics to inform future moments of uncertainty. Second, this chapter spotlights the intersections between health and transportation. It builds on the preliminary analysis of Covid and mobilities (Huang and Kwan 2021; Kar, Le, and Miller 2021) as well as how Covid affected the transportation sector in South Africa (Luke 2020) and

DOI: 10.4324/9780429322280-8

elsewhere on the continent (Mogaji 2020). This chapter adds to these discussions by bringing the broader perspective of transportation failure into contemporary reflections on mobility and transportation restrictions.

The next section grounds this chapter in a broader reflection on public health and cities. It contextualizes the current pandemic within a long history of disease. From there, this chapter considers disease in South African cities, in particular reflecting on how past experiences with bubonic plague, Spanish flu, and HIV/AIDS impacted urban governance and planning. This chapter then refines its focus to mobilities and Covid. It applies these to discussions in South Africa: first, I reflect on the role of all forms of mobility restrictions, second, those on international transportation, and third, those on public transportation. In conclusion, this chapter reflects on the intersections between health, transportation, and urban geographies.

Public Health, Epidemiology, and Urbanization

The first case of Covid was recorded in Wuhan, China on December 31, 2019; on January 13, 2020, the first case outside of China was confirmed in Thailand; and on March 11, 2020, after it spread to every corner of the globe, the WHO declared it a pandemic (WHO 2020b). As of mid-2025, there have been over 700 million confirmed cases of Covid and more than 7 million confirmed Covid-related deaths (Worldometer 2024b). Scholars have responded to these staggering figures by reflecting on the overlaps between Covid and the urban economy (Florida, Rodríguez-Pose, and Storper 2021), urban infrastructure (Connolly, Keil, and Ali 2021; Honey-Rosés et al. 2020; McFarlane 2021), and urban governance (Enright and Ward 2021; Kleinman 2020; Parnell 2020) in a post-pandemic world. Much of these commentaries and thought pieces are appropriately speculative and cautious given the circumstances. Research on disease and the city did precede the pandemic (Ali and Keil 2007, 2008; Wolf 2016). However, the pandemic has stimulated both the need and opportunity to reflect on the intersection between disease and mobility.

Disease, pandemics, and health crises are not new to cities: the Great Plague destroyed Athens during the Peloponnesian War in 430 BC (Kelaidis 2020); a millennium later, bubonic plague killed one-third of Europe's population (Benedictow 2013); and water-borne illnesses like cholera and typhoid were endemic in London, Paris, and New York until this past century (Glaeser 2020). The 1918–1920 Spanish flu killed 50 million worldwide and as many as 675,000 in the United States alone (Correia, Luck, and Verner 2020). Fears of new and re-emerging diseases, including HIV/AIDS (Phillips 2001), Ebola (Wilkinson and Leach 2015), and Severe Acute Respiratory Syndrome (SARS) (Ali and Keil 2007, 2008) have been at the forefront of public health concerns for decades.

Likewise, concern for public health in the city is not a new field. The journalist Jacob Riis published a book of photographs documenting the dirty and overcrowded slums of New York City in "How the Other Half Lives" (1890). Similar first-hand accounts document the deprivation of the working classes in London's

East End (London 1903) and in England's industrial north (Orwell 1937). The emergence of the modern city, as opposed to the "bacteriological city," is associated with the development of modern infrastructure (Gandy 2004). Over the next half-century, living conditions steadily improved throughout the global north, yet dangerous conditions persisted in the rapidly urbanizing cities of the global south. In response, the WHO launched the Healthy Cities Project in 1978 to encourage local authorities to engage with issues of health and well-being (Barton et al. 2015). Over the next two decades, "healthy urban planning" (Barton and Tsourou 2000), as it was dubbed, was promoted to braid together the principles of health and urban planning. Today, there are more than 2,000 municipalities linked to this initiative (Barton et al. 2015).

What makes this moment unique is the speed and frequency with which zoonotic diseases – such as influenza, tuberculosis, and Ebola, which are passed from animals to humans, as opposed to vector-borne diseases such as malaria and dengue fever, which are spread by mosquitos and other parasites – are emerging. Vector-borne diseases are usually confined to specific climatic conditions and spread slowly, while zoonotic diseases are not geographically contained and therefore spread quickly (Wolf 2016). Zoonotic diseases are therefore more likely to be increasingly prevalent in cities around the world. In 2014, epidemiologists were surprised when Ebola, having remained a rural illness for nearly four decades (Wolf 2016), showed up in Lagos; and the outbreak of SARS in China and its rapid dispersal as far as Toronto (Ali and Keil 2007) indicates a shifting relationship between urban health and disease.

The urbanization of Ebola and SARS, among other diseases, has led to a suspected association between population density and disease. Kendall and others (1991), for example, claim a connection between the spread of infectious disease and the urban built environment. Drawing on cases of dengue fever in Mexico and Honduras, the authors blame stagnant water in the slums for breeding disease-carrying mosquitos. More recently, Wilkinson and Leach (2015) argue that the density in Monrovia and Freetown fueled the spread of Ebola.

In the recent pandemic, as in ancient ones, there were concerns that Covid flourished in denser urban environments where overcrowding and poor air circulation increased vulnerability (Choi and Denice 2022). This was compounded by the high rates of infection in cities – roughly 90 percent of Covid infections worldwide were reported in urban settings (United Nations 2020). Hamidi, Sabouri, and Ewing (2020) were early to ask whether density aggravates the pandemic. In their analysis, they discovered that "connectivity matters more than density" in the spread of Covid. The movement of people between and within cities and countries, rather than their density, was a significant factor in the pandemic. Florida, Rodríguez-Pose, and Storper (2021) similarly note that in the first wave of the pandemic, those cities with better connectivity – as measured by high numbers of international airports and tourists – were harder hit. But they also observed that in subsequent waves, small cities and rural areas had higher case numbers per capita than big cities. In their analysis of the spatiality of disease, Connolly, Keil, and Ali (2021) also argue against any association between density and Covid. Instead, they note how urban sprawl, with its unsustainable features, is just as likely to feed outbreaks. Cities, they reason, tend to have the resources that enable quicker, more effective containment methods.

Indeed, rather than densities in cities, it is their inequalities – unequal access to housing, transportation, jobs, and services – that could serve to fuel the rapid distribution of disease. In studying case numbers in New York, Florida, Rodríguez-Pose, and Storper (2021) demonstrate that zip codes with the lowest income per capita experienced the highest incidence of Covid. Whereas knowledge workers were generally able to work from home, service workers – cleaners, bus drivers, and healthcare workers – had to go in to work and this typically necessitated travel by public transportation. Just as transportation is experienced differently by different populations, health – or lack thereof – highlights the economic disparities of urban life.

Disease and the South African City

South Africa is no stranger to disease, having been hit by smallpox between 1713 and 1893, bubonic plague from 1901 to 1907, poliomyelitis from 1918 to 1963, and HIV/AIDS from 1982 through the current era (Harrison 2020). However, none of these antecedents were as challenging as the Spanish flu, a particularly virulent strain of the H1N1 influenza virus, which first arrived in Cape Town in September 1918 and then traveled to Johannesburg the following month. The *Rand Daily Mail*, South Africa's largest newspaper at the time, reports the arrival of the flu as follows:

> Spanish influenza which has been so prominent in England, has now made its appearance in South Africa, a few natives who recently returned from overseas being reported to be suffering from the complaint. The strictest precautions have been taken to prevent the spread of the disease.
>
> (Rand Daily Mail 1918)

While the media attributed its first arrival to the returning soldiers of the South African Native Core, evidence suggests that the Spanish flu was already present in Durban via soldiers who had come back from the United States. In what became known as "Black October," the Spanish flu killed 6 percent of the population – 300,000 South Africans – in just six weeks (Webster 2020a, 2020b).

South Africa responded to the devastation of the Spanish flu through changes in governance and urban planning. First, when the Union of South Africa was formed in 1910, no provision was made to systematize public health. Instead, municipal and provincial authorities retained responsibility. However, their limited resources quickly proved ineffective, and in 1919, the national government passed the Public Health Act, which established the first national Department of Public Health (Phillips 1987).

Second, the Spanish flu enabled the institutionalization of segregated urban planning and later the passage of the Native (Urban Areas) Act in 1923. Emphasis was placed on preventative rather than curative medicine explains Phillips (1987). Indeed, this was not the first instance of Swanson's (1977) "sanitation syndrome," a European phobia that attributes racial segregation to a moral panic associated with poverty and disease, which emerged in the late 19th and early 20th centuries. The outbreak of bubonic plague between 1901 and 1904 was largely blamed on the native population: in King Williams' Town, the magistrate insisted that the natives had spread smallpox even though it was obvious that it had arrived with rats aboard a train from a ship docked in East London; and in Durban, officials used similar racialized language even when it was clear that the outbreak stemmed from the harbor (Phillips 1988). The ensuing panic led to the forced relocation of Blacks from Cape Town to Ndabeni and from Johannesburg to Klipspruit (Swanson 1977). "Panic" was a "metaphor to shape perceptions and influence or justify behavior" that was used to implement spatial divisions (Swanson 1977, 388). For Maylam (1995, 25), the "metaphorical significance of syndrome" came to replace the reality of the disease. Similar outbreaks occurred in Harare and Nairobi as well as in the Democratic Republic of Congo, Morocco, and Senegal, with municipalities also encouraging segregation patterns (Njoh 2007).

Movement has long been considered a key mode of transmission of diseases in South Africa (Phillips 2012). In the 18th century, smallpox traveled from Dutch colonies in the clothes of passengers and crew who developed the disease during the voyage to Cape Town. In 1812 and 1840, the disease arrived abroad on a slave ship on its way from Mozambique to Brazil, and in 1882, it was brought to Cape Town on a steamer from Britain. It then traveled on the newly built trains throughout the region. In 1883 and 1893, outbreaks in Kimberley and Johannesburg were later traced to the new rail connections with Mozambique. Phillips (1988) argues that it is the migrant labor system – not mobility – and the newly developed transportation networks that were central to the transmission. In the 18th and 19th centuries, smallpox was transported via sailors and migrant laborers; in 1918, Spanish flu via discharged Union Defense Force soldiers and homeward fleeing mineworkers; in the 1920s and 1930s, typhus and Asian flu via newly recruited mineworkers; and in the 1980s, HIV/AIDS came with mineworkers, truck-drivers, and returning soldiers. For Phillips (2001), these epidemics all share a similar characteristic – young men as the carrier of disease; and I would add, on their way to or coming from being exploited by the capitalist system that forces them to move.

Transportation and COVID-19

Transportation and mobility restrictions were intrinsic to managing the Covid pandemic (Glaeser, Gorback, and Redding 2020; Harris 2020). Working from home, e-learning, and restrictions on public gatherings led to major reductions in both public and private travel. Moreover, shifts from public to private modes of travel were widespread, especially as people avoided busy buses and trains during peak travel hours. The intended goal was to reduce social interactions by lessening the density defined as "as numbers moving through space, including transportation systems, streets and the in-between spaces of city-centre shopping" (McFarlane 2021).

Indeed, mobility has always been integral to the spread of infectious diseases. Bubonic plague, caused by a zoonotic bacterium *Yersinia pestis*, first moved from China to Crimea via the traders along the Silk Road. From Kaffa, a seaport on the

Black Sea, it is thought that the disease traveled to Constantinople and Athens before making its way to Genoa, Sicily, and Venice (Yue, Lee, and Wu 2017). The transmission was particularly mobile, according to Welford (2018), for two reasons: first, traders moved back and forth between caravanserais rather than traveling the entire route, which meant more exposure to more people; and second, the Silk Road included multiple branches, which expanded the topographical distribution of the disease, eventually spreading it to cities in Egypt, Iraq, and Persia. Bubonic plague later spread to the Americas through the expansion of Spanish colonialism along with smallpox and measles, killing 90 percent of the local population; cholera is thought to have originated in Russia but it was infected British soldiers who brought it to India; and the 1918 Spanish flu spread through postwar movement of troops and goods.

During the pandemic, mobility and transportation restrictions took on three different and overlapping forms: in some instances, all mobility was restricted. In December 2020, China was the first country to introduce Covid quarantines, only allowing essential trips to the grocery store and the doctor. This concept quickly spread throughout South East Asia and to France, Italy, and Spain, the first European countries to enact similar measures (Kar, Le, and Miller 2021). However, the concept of quarantine dates back to 1377 when Dubrovnik introduced maritime quarantines (Tognotti 2013). In 1423, Venice banned sailors from entering the city for 40 days – and from that the word quarantine is derived from the Italian *quaranta giorni*, which translates to "40 days" (Mockaitis 2020). The Venetians established plague hospitals – "lazarettos" – in which officials sealed infected people in their homes or hospitals specifically established for the infected (Welford 2018). They also introduced procedures for "purgation" of the products most likely to transmit disease – such as cloth, leather, and wool. These quarantine procedures spread thereafter to Genoa in 1467 and Marseille in 1476 to quell Bubonic Plague and remained the practice in London in 1663, New York 1688, and Boston in 1691 to stop the spread of Small Pox (Tognotti 2013).

Restrictions on international travel between countries continued into the 21st century. These approaches were based on previous attempts to stop pandemics by preventing international movement and migration – none of which were substantively successful at preventing the spread of disease. In the case of Ebola in Lagos, a traveler from Liberia is said to have been ill when boarding a flight; and SARS is believed to have spread by a businessman traveling from Guangdong, where the disease first originated; to Hong Kong, where the disease spread (Wolf 2016). In each of these instances, the real and imagined fears of the outside/r remained the common approach – even though each time the disease ultimately reached beyond its point of origin. These movements expose the character of "networked disease," a term developed by Ali and Keil (2008) to explain the complex relationships between globalization, cities, and disease. "As the scales hierarchy of global cities becomes the conduit of disease transmission, another reality becomes visible," the authors suggest. "[T]he global cities hierarchy is really a complex network of topological relations both externally (cities among one another) and internally (the capillary system of the globalized metropolis)" (2008, 5–6). The WHO suggests that the recent spread of Ebola might be associated with the increasingly high degree of movement across Africa – the incidence of migration and movement is seven times higher in Africa than anywhere else (WHO 2015). Ironically, the undisturbed landscapes and unpaved roads of the Democratic Republic of the Congo may have limited the spread of Ebola in the past. But upgrades to transportation infrastructure – for example, the paving of a 550-kilometer road connecting Kikwit and Kinshasa, which reduced travel time from a week to just eight hours – is accelerating the spread of infectious disease (Yong 2018). In short, today's epidemics are firmly embedded in cities and networks that exceed their geographies.

By the end of February 2020, 38 countries had reported travel bans to and from China, denying entry to travelers, restricting visas, and mandating quarantines on arrival. Some countries suspended all flights. Soon after the WHO advised against travel bans explaining that "restricting the movement of people and goods during public health emergencies is ineffective in most situations and may divert resources from other interventions." Restrictions interrupt business as well as aid and "may have negative social and economic effects on the affected countries" (WHO 2020a). In 2021, the WHO amended their instructions explaining that travel bans disincentivize countries from sharing their epidemiological data (WHO 2021). Instead, the WHO recommended that countries enhance their testing and monitor the variant's behavior. They likewise advised vaccinations, hand hygiene, masks, social distancing, and ventilation.

Third, usage of public transportation was restricted. Both because of these lockdowns and due to general concern, the Covid pandemic led to substantial shifts in the supply and demand for urban public transportation. Starting in March 2020, urban transportation ridership plummeted around the world – Istanbul lost 50 percent (more than two million) riders in the first three weeks of March 2020 (Null and Smith 2020); on March 24, ridership on New York's subway fell 87 percent, or nearly 4.8 million riders (Goldbaum 2020); to maintain social distancing in May 2020, only 50,000 passengers were permitted to board London's Underground system every 15 minutes – as opposed to the pre-pandemic normal of 325,000 (Butcher and Milne 2020). In several cities, such as Wuhan, China, public transportation was suspended entirely (Null and Smith 2020). In the aftermath of these restrictions, cities created ambitious programs to re-encourage public transportation usage. Disinfectants and hand-washing stations were installed in bus systems from Istanbul to Kigali; and in many cities, bus passengers started boarding from the backdoor and no payment was collected on board (Null and Smith 2020). Several cities also responded by encouraging cycling: New York created a "Bike or walk to work if you can" campaign; Milan added 35 kilometers of new bike lanes (Honey-Rosés et al. 2020); and Bogotá expanded the city's ciclovia program by 76 kilometers. Meanwhile, other cities took strict action against cyclists: Spanish cities reportedly fined cyclists for taking non-essential trips during lockdown (Null and Smith 2020). Cities also announced temporary and permanent changes to pedestrian infrastructure. In global south cities, however, where sidewalks tend to be more heavily used and crowded, social distancing measures were more difficult to implement. In Indian cities, for instance, where the sidewalks are frequently occupied by informal traders, the government marked the sidewalks with yellow circles encouraging sidewalk-users to keep a distance (Honey-Rosés et al. 2020).

The evidence above suggests that mobility inequalities are exacerbated in moments of crisis. For urban transportation, only those with fewer financial opportunities continue to rely on buses and trains, while those with a little more money who may have previously also used public transportation turn to private transportation. While many office workers can work from home, many others are on the front lines risking their health. Hamidi, Sabouri, and Ewing (2020) note that US counties with a higher percentage of educated individuals have lower rates of Covid infection. They attribute this finding to the fact that these workers are typically able to work remotely and are therefore less likely to use public transportation to travel to work. By contrast, residents in low-income neighborhoods are less likely to work remotely and more likely to rely on public transportation. In thinking beyond the pandemic, it is important to realize that transportation systems should be egalitarian and democratic as well as sustainable.

This is not an exhaustive discussion of the associations between mobilities and disease in the current pandemic. There have been shifts in trade and maritime transportation, in particular affecting the global distribution of essential commodities like food. However, this section is meant to draw attention to the need for further deliberations on mobilities and Covid, an approach that I will now take by grounding the remainder of this chapter in South Africa.

Impact of COVID-19 on Transportation in South Africa

Restrictions on mobility and transportation have been one of the primary approaches taken by South Africa to quell the pandemic. Drawing on the discussion of transportation and Covid, this section now focuses on South Africa's experiences in the pandemic. In particular, I reflect on three types of restrictions – on mobility, on international transportation, and on urban transportation. Together, these three reflections provide a critical understanding of how health and transportation intersect and intertwine as well as why transportation failure matters in South Africa.

Restricting Mobility

On March 5, 2020, the South African Department of Health confirmed the first case of Covid: a South African tested positive upon returning to Durban from Milan (Maja 2020). On March 15, President Ramaphosa declared a national state of disaster. Between March 27 and April 16, South Africa had a 21-day national lockdown that included directives to work from home, banning the sale of alcohol, and suspending public transportation (BBC News 2020).

The effect of these lockdown measures on South Africa's already floundering economy cannot be overstated: GDP fell by over 16 percent in the first and second quarters of 2020, resulting in an annualized growth rate of – 51 percent. Household expenditure decreased by nearly half; and spending at restaurants and hotels ceased almost entirely (Stats SA 2021a). In 2020, eight of South Africa's largest industries recorded economic losses: construction (–20 percent), manufacturing (–12 percent), mining (–11 percent), and trade (–9 percent), contributing to a total annual contraction of 7 percent of GDP. According to the South African Reserve Bank, this is the largest decline in GDP since record-keeping began in 1960, with the second largest contraction recorded in 1991 at 2.1 percent. In real terms, the economy in 2021 is about the same size as it was in 2012 (Stats SA 2021a).

The extensiveness of these restrictions on mobility has led to employment losses. Three million South Africans lost their jobs in the first month of lockdown, and 21 percent of households went hungry (Haffajee 2020). In December 2020, the Department of Small Business Development reported that 42.7 percent of small businesses had closed (Business Tech 2020). And, by December 2021, the official unemployment rate in South Africa reached 35 percent, which amounts to 7.6 million unemployed people with more than 660,000 people losing their jobs in the third quarter alone (Stats SA 2021b). Accordingly, there has been tremendous criticism of the government's response – the unknown duration of lockdowns, the erosion of civil liberties, the presence of the military, night curfews, together with bans on outdoor exercise and the sale of alcohol.

Restricting International Transportation

The practice of restricting international travel has been another common response in South Africa. Through the Covid pandemic, South Africa implemented a range of restrictions on international entry. After the initial policy of 21 days of quarantine, South Africa followed a phased return: under National Alert Level 4, international travel into South Africa was only permitted via the major airports in Johannesburg, eThekwini, Cape Town, and Kruger Park. All travelers were required to provide proof of a negative Covid test obtained within 72 hours of departure. Travelers were also required to test upon arrival, and if they tested positive they were required to quarantine for ten days. Under levels 1–3, international travel was permitted, many of South Africa's 53 land borders reopened, and long-distance bus and rail travel (more than 200 kilometers) was allowed (South African government 2022).

Whereas South Africa's border remained open for much of the pandemic, its citizens were not permitted to travel into many countries. For example on December 24, 2020, just a few months after removing the travel ban, the United Kingdom announced an immediate travel ban on South Africa. This followed three previous rounds of travel restrictions, each of which devastated the economy. In reflecting on the spread of Ebola, Yong (2018) suggests that flight bans may increase the disease transmission by driving patients to seek alternative, illegal transportation options. Likewise, it discourages health workers from helping contain foreign outbreaks for fear that they will not be able to return home.

International travel bans are particularly detrimental to the tourism industry. It is estimated that globally in 2020 travel restrictions cost US$4.5 trillion and cost millions of jobs. Across the African continent, travel bans caused a 49.2 percent decline in GDP. In South Africa, travel restrictions caused a 96 percent drop in tourist visits to national parks, which equates to 90 percent of total tourism revenue for the country. Canceled hunting trips and reductions in the sale of live animals and meat products resulted in US$406 million in losses (Braczkowski and Biggs 2021). After this round of international travel, bans were announced, the South African tourism industry, which contributes R82 billion annually and creates an estimated 1.5 million jobs, was devastated (Makhaye and Mkhize 2021).

Restricting Urban Transportation

Urban public transportation was severely affected during Covid restrictions which simultaneously reduced both the supply and the demand for public transportation. South African cities adopted measures to reduce passenger loads, the frequency, and duration, of trips. Several cities suspended public transportation services altogether. This made it challenging for essential workers to reach their destinations (Diouf et al. 2020). Likewise, cities noted a decline in demand for public transportation through national lockdowns. In South Africa, trip demand was reduced by as much as 80 percent during the 21-day lockdown (Diouf et al. 2020). This reflection on urban transportation in Covid provides an important glimpse into the overlaps between mobility and disease.

Concern for the transmissibility of Covid on public transportation also drove shifts in the modal split of public transportation. For the most part, bus and rail services were better able to comply with health and hygiene requirements. For example, the highly regulated Gautrain rail service that connects Johannesburg and Tshwane was a glowing example of health and hygiene. Hand sanitizers were installed in all stations and communications from the Department of Health were broadcast on the internal TV stations. All touchpoints were washed and sanitized regularly (Pikoli 2020). To better understand the risks to both drivers and passengers, Stent, Gontsana, and Washinyira (2021) measured the transmissibility of Covid in Cape Town on different modes of public transportation – trains, buses, minibus taxis, and Amaphelas. Using carbon dioxide (CO_2) as a measure of transmissibility in each vehicle, they recorded the lowest CO_2 emissions on trains (mean: 1,000 parts per million, or ppm), somewhat higher levels on buses (mean: 1,150 ppm), and the highest levels in taxis (mean: 1,800 ppm). This suggests that public transportation usage may not be as unsafe as presumed. While this study is "journalistic, rather than scientific" and authors caution "strict inferences from our data are ill-advised," their attempt to quantify the role of public transportation in the transmission is noteworthy.

The effects of the pandemic on the minibus taxi industry cannot be overemphasized. Whereas, on a typical day in South Africa, 65 percent of trips are made by minibus taxi, 20 percent by bus, and 15 percent by train (Mndebele 2019), during the pandemic public transportation usage changed dramatically. The irregular nature of the taxi industry makes it particularly vulnerable to disease transmission. And when the industry was shut down in March 2020, the reverberations were felt across the country. Taxi drivers, queue marshals, and car washers, as well as the informal traders employed directly and indirectly by the industry, were out of work. Likewise, commuters were left stranded (Nkosi 2020). Determining how best to manage and support the underregulated, politically powerful taxi industry during the pandemic has been a challenge. During the 21-day national lockdown, the government reduced the number of minibus taxis in use, first to 50 percent and later to 70 percent of normal capacity. In reflecting on the rationale for lifting the restrictions in April 2020, Transport Minister Fikile Mbalula explained,

> After our announcement of catering for 100%, there was a public outcry about the impact of such an arrangement to the health safety of public transportation users. We took these issues seriously and engaged with key stakeholders, resulting in a consensus of maximizing the loading capacity to 70%.
> (Simelane 2020)

To resume operations, both the driver and the passengers were required to wear N95 masks. The costs of such protective equipment, however, are substantial – around R2,000 for a box of 10, excluding VAT – making compliance difficult (Simelane 2020). There has also been tremendous controversy around if, and how, to provide appropriate economic relief to the taxi industry. In January 2022, after months of back-and-forth negotiation between the national Department of Transport and Santaco, the main representative of the taxi industry, a R1.1 billion relief fund was announced. This equates to about R5,000 per taxi. Santaco by contrast had requested R20,000 per taxi if they were to continue operating at only 70 percent capacity. To be eligible for the Taxi Relief Funds, taxi drivers must be South African citizens or permanent residents, in possession of a valid operating license, and registered with the South African Revenue Service (Venkile 2021).

South Africa's urban rail has also experienced recent hardships. Before March 2020, rail moved about two million people a day in Gauteng, KwaZulu-Natal, the Eastern Cape, and the Western Cape. Trains might be preferred over taxis or buses because they are a cheap and comfortable option. However, the rolling stock is old and poorly maintained, there is little security, and the trains are often overcrowded. When the trains were closed in March 2020, the stations were vandalized. Throughout the system, the iron tracks were stolen, the high-voltage electricity cables cut, and the station infrastructure was wrecked. Some of this vandalism pre-dates the pandemic but in the absence of security or commuters, conditions deteriorated dramatically (Nkosi 2020). Passenger Rail Agency of South Africa, the state-owned enterprise responsible for operating South Africa's trains, has projected a loss of nearly R200 million for every month that it was not operating (Payne 2020). To resume services, they were required to implement screening and testing at stations, deep cleaning and disinfection at all depots and stations, disinfection of all trains after each trip, and issue of protective equipment to all employees (Payne 2020).

Together, these reflections demonstrate the unevenness of transportation-fueled Covid restrictions. It has not been my intention to criticize government measures, nor to assess the epidemiological merits of lockdowns; rather this chapter exposes how these measures exacerbate the inequality in South Africa – where some work is able to easily shift online, whereas other work is stopped, and following those patterns, some transportation modes are better able to adapt to the necessary shifts in economic and social life. This intricate look at the transportation sector exposes how lives and livelihoods have experienced Covid differently.

Conclusion

This chapter engages with South Africa's attempt to stem the spread of Covid by introducing restrictions on mobility, on international transportation and on urban public transportation. It draws on South Africa's long history of pandemics and disease to explain the government's decisions to restrict mobility during the Covid pandemic. Indeed, over the past two decades, South Africa has developed one of the best methods for molecular testing in the world. They use this same technique for HIV viral load testing and South Africa was therefore much better prepared for Covid testing than most countries (Malan 2021). The same is true in Liberia and Sierra Leone – because of the recent experience with Ebola they were the first African countries to implement a quarantine as well as other cautionary measures (Diouf et al. 2020). Restricting mobilities alone, however, did not stop the spread of Covid. Instead these extensive restrictions on mobility led to considerable economic and social challenges in South Africa as well as elsewhere and likewise these measures failed to protect four million South Africans from contracting the virus and 100,000 from dying from it. Rather, it was protective measures, testing, and research – like those which were done in South Africa to sequence the Omicron variant – that actually helped.

For geographies more generally, it is far beyond the intention of this chapter to make any future predictions about the post-Covid city or draw any conclusions on the efficacy of mobility-focused responses. What will be the long-term changes to the transportation sector? In what ways will public transportation usage change? To what extent will (would-be) public transportation users be nervous, instead opting to travel by private vehicle? Or, will there be, as Florida, Rodríguez-Pose, and Storper (2021) so eloquently point out, "social scarring" after the pandemic? Is there such a thing as a "post-Covid city" (Florida, Rodríguez-Pose, and Storper 2021; Shearmur et al. 2021), and how can the transportation sector prepare for this eventuality? As this particular pandemic winds down, I could only speculate. Rather, it has been my intention to open up a space for further problematization and reflection on what happens when transportation fails.

For the broader discussion of the book, this chapter has aimed to highlight the intersections between health and transportation especially as it relates to transportation failures. By drawing on the history of diseases like bubonic plague, Spanish flu, and HIV/AIDS, this chapter has demonstrated that transportation failures are not merely theoretical concerns – indeed when transportation fails, we all suffer; and because we have not adequately considered the intersections between health and mobilities, we were not prepared for the Covid pandemic. Going forward, we need to better understand conceptual and practical intersections between health, transportation, and urban geographies. These analyses are essential for discussions of climate change, mitigation, and resilience – concerns fundamental to the shifting nature of our cities and our planet. Transportation failures therefore are not only urban concerns but planetary concerns – reflections that continue in the final conclusion chapter.

9 Conclusion

Introduction

Browsing through my children's bookshelf – *Bus Stops* by Taro Gomi (1988), *Things that Go* by Stephan Lomp (2017), and *That's Not My Bus* by Fiona Watt (2020) – I see how our hopes and dreams for transportation – that it fuels social cohesion, economic growth, and urban politics – are dismissed, discounted, and discontinued as we grow up.

Returning to December 2023 when Johannesburg's Rea Vaya bus rapid transit (BRT) operator declared bankruptcy. Its creditors attempted to seize its fleet of buses and commuters wondered how they would get to and from work (Motsoere 2023). But, this was not the first frustration and failure for transportation in South Africa – far from it. In 2019, the "City of Joburg condemns vandalism of Rea Vaya station" (Zulu 2019), and a few years later, "Rea Vaya has no answers for Thursday's temporary suspension of services" (Gounden 2023). I texted a friend, "There aren't enough emojis to express my disappointment," to which she promptly responded, "We might need a support group to deal with the sadness." This book attempts to make sense of transportation failure by contextualizing it within a range of burgeoning geographical literature.

This concluding chapter summarizes the arguments made in this book and reflects on why transportation fails in South Africa and beyond. In the next section, I outline my main contributions to the literature, draw connections between the chapters, and return to the main aims. From there, I reflect on transportation in South Africa in particular amidst the complex economic, social, and political landscape in South African cities. Finally, I expand the focus beyond South Africa to contemplate why transportation fails in both the global north and global south.

Reflecting on Why Transportation Fails

This book brings failure to the forefront of transportation and urban geography. It draws conceptual richness from a range of literature and expands both theoretically and practically to answer this question of why. In doing so, it makes two overarching contributions – first to further critical analyses of urban transportation in Africa and second to center transportation justice within these discussions.

DOI: 10.4324/9780429322280-9

The chapters draw from a range of scholarship, including decolonial geographies (Schwanen 2018a; Wood, Kębłowski, and Tuvikene 2020), Black geographies (Alderman, Williams, and Bottone 2022; Inwood, Alderman, and Williams 2015), policy mobilities (McCann and Ward 2011; Wood 2022a), walking geographies (Behrens 2005; Zavetoski and Agyeman 2015), animal geographies (Wolch 2002; Wolch, West, and Gaines 1995), art geographies (Bissell and Fuller 2017; Hawkins 2019) and health geographies (Ali and Keil 2007, 2008; Wolf 2016). Throughout, I argue that transportation fails because it overlooks these disciplinary discussions, adopting a relatively straightforward and linear understanding of failure rather than a complex and pluralistic one. This incomplete understanding is why I argue transportation is failing in South Africa and across the globe. Each of the chapters draws out and highlights ways in which economic, political, and social creativity; experimentation; and entrepreneurialism have been overlooked by transportation planners with significant consequences.

Chapter 2 "Decolonizing Transportation" and Chapter 3 "Black Transportation" draw on burgeoning literature in decolonial and Black geographies to demonstrate both the need and the opportunity for drawing from a wider theoretical register. These two chapters contextualize contemporary conversations providing critical reflections from within feminist, Indigenous, and queer literatures. These frameworks are particularly important for transportation geography, which has traditionally been developed by exclusionary systems. The chapters also rely on historical evidence to rethink contemporary conditions. They then explore processes of decolonizing transportation – by reinterpreting knowledge from elsewhere (e.g., transit-oriented development), rejecting knowledge from elsewhere (e.g., bike-share schemes), and refining local knowledge (e.g., minibus taxi) – and Black transportation – through the (White) policies and laws that control movement, community action, and protest against racist transportation and planning for the formalization of the minibus taxi industry. Both chapters use the minibus taxi industry as an example to critically reexamine what it means for transportation to fail, the ways in which transportation failure can be blurry, and the extent to which failure is subjective.

Chapter 4 "People-Oriented Transportation" and Chapter 5 "Walking-As-Transportation" discuss the transportation concepts of transit-oriented development and walkability. These two chapters use popular transportation models to rethink transportation failure. Both models have been popularized around the world and introduced into South Africa as ostensible successes. However, they have so far failed in situ. Both chapters discuss Johannesburg's Corridors of Freedom, a city-wide plan for redesigning both transportation systems as well as the surrounding economic and spatial amenities. They note the failures of this plan to transform transportation in South Africa and illustrate the overlaps between transportation failures and urban failures.

Chapter 6 "Trans-paw-tation" and Chapter 7 "Transp-art-ation" extend the discussion into animal geographies and art geographies. While "Trans-paw-tation" reflects on the movement of animals and animals' movements in both the historic and contemporary city, "Transp-art-tation" explores the decorations adorning bus and train stations, the art on transportation vehicles, and the abstractions that narrate our movement. Both chapters note the ways in which the (human and animal) public are excluded from transportation and the implications of this disregard on transportation failure. Informality plays a central role in these discussions. In so doing, these narratives also further the complex conversations in social and urban geographies and demonstrate how a transportation lens can contribute to both theoretical approaches.

Chapter 8 "When Transportation Fails" updates the discussion in the aftermath of the Covid pandemic. It reflects on what happens when transportation fails by analyzing the failings that happened not only in the recent Covid pandemic but also in past pandemics. This chapter is essential for reminding the reader that these theoretical challenges become real emergencies when weak transportation systems are not strengthened.

The categorizations in these chapters are not rigid. Indeed, the empirical stories of transportation in South Africa could move across the chapters – much as transportation policies and systems do in and between cities. Likewise, there are many contradictions evident across the chapters – the extensive investment in transportation infrastructure versus a laissez-faire approach to the minibus taxi industry; or, more specifically in Johannesburg, government-led planning for the Corridors of Freedom versus community-inspired planning for walkability, to name just a few. These can be read as misalignments between the state and residents. They constitute ostensible incongruities that reveal the potential potency of transportation to cut across the city, and if supported, merge across economic, political, and social divisions.

Despite the title, in many ways, this book is not about the technicalities of transportation failure. And, perhaps too much has already been written about transportation disappointments and disasters, especially in Africa: the highest rate of road fatalities in the world, the highest transportation costs in the world, governments struggling to cover the costs of transportation investment, while informal systems provide unreliable and dangerous services for the poor and the marginalized. Instead, this book provides an opportunity to reframe and reinterpret transportation failure using a host of alternative lens. In so doing, it highlights art and architecture, more-than-human animals and more-than-animal humans, to expose the myriad of ways by which transportation does not only crash and collapse.

Nor is it my intention to present success/failure as straightforward, objective, or permanent binaries. Indeed, transportation failure in South Africa is not always obvious – just like it is not always easy to distinguish between formal and informal modes of transportation, or public and private transport. There are instances where informal transportation is successful – for some low-income users it provides a much-needed, affordable mode of transportation between their destinations, and it does so without any direct cost to government; while to others, it is dangerous and chaotic, and costs the public by remaining untaxed. In this book, we learn that success/failure is a relative concept with gradients of success/failure blended between, across, and throughout. In some instances, it is a matter of success/failure for whom?; some may recognize success even while others perceive failure. It all depends on where you stand or what bus you ride. Rather, my objective has been to use this seemingly far-reaching literature to open up space for alternative thinking about potential solutions. In the next section, we return to discuss transportation in South Africa and how we might move beyond failures.

Reflecting on Transportation in South Africa…

By most measures, South Africa's public transportation systems are failing. Bus and rail ridership are declining while both systems are already overly subsidized; rail infrastructure is deteriorating and attempts at the procurement of rolling stock have led to expenditure on incompatible train cars; while the bold experiment with BRT is incomplete and inadequate. It's no exaggeration to say that South Africa's transportation has reached "day zero." This book provides a written account of the postapartheid efforts to reform transportation. It features two decades of postapartheid attempts at transformation, which despite its failures remain vital for South Africa's postapartheid future. This story of failure, however, is not unique to transport. Indeed, transportation is just one of many failures of the postapartheid dream of social and spatial reconciliation.

Housing, for example, has also failed either to materialize or to make substantive inroads toward postapartheid change. When the African National Congress (ANC) took power in South Africa's first fully enfranchised elections in 1994, the new government promised comfortable and safe homes for everyone. The 1996 national constitution went even further, making housing a fundamental human right and enshrining housing need into law. Between 1994 and 2022, the government built more than three million homes and provided about 12.5 million residents with ownership titles for no, or low costs (Orderson 2024). However, more than 30 years later, over 3.4 million people in South Africa remain on the waiting list for a home (Cele and Siwele 2024). In Johannesburg alone, more than 400,000 families are on the list for a council house (Cele and Siwele 2024), and that list has grown to more than 600,000 in the Western Cape (Orderson 2024).

The un-housed have few other options other than to live in informal settlements on the periphery of the city or hijacked buildings in the inner city. In Cape Town, for example, no new affordable housing has been built in the inner city in the three decades since the end of apartheid. Residents therefore developed a system of "backyard dwellers," whereby an inexpensive sublet is created on part of their property to other even more poverty-stricken residents who erect a makeshift dwelling, typically made of corrugated tin and lacking basic electricity, water, and sanitation services (Lemanski 2009). Elsewhere people rely on hijacked buildings – office towers, apartment buildings, and factories abandoned by their owners and captured by criminal syndicates who force residents to pay rent and protection fees in exchange for slum-like conditions (Wilhelm-Solomon 2020). In Johannesburg, there are more than 600 hijacked buildings in the inner city alone (Cele and Siwele 2024), and in August 2023, 70 people died when their hijacked building caught fire without adequate fire alarms or other protections. One of the reasons for the growth of hijacked buildings is their central location in the inner city close to economic opportunities. Indeed, the peripheral location of government-built housing is one of the biggest complaints after availability (Cele and Siwele 2024). The location of these developments on the outskirts of town, far from economic and social opportunities, has kept even the born-free population excluded from urban life and dependent on informal, expensive, and unreliable modes of transportation.

The story repeats itself with regard to energy supply, where rolling blackouts have plagued South Africa for nearly two decades. Eksom, the sole public utility provider in South Africa, can supply just 27,000 megawatts of the winter peak demand of 37,000 megawatts and a summer peak demand of 32,000 megawatts (Farmer 2023). As a result, load-shedding – a scheduled system of blackouts when the energy demand exceeds production to prevent the power grid from overloading – has become commonplace. In 2023, households in South Africa lost power for 280 days for up to 12 hours, an increase from 65 days in 2022 (Rose 2024). Much of the reason for energy failures can be attributed to Eskom, which has struggled for decades with corruption and mismanagement, but some can also be blamed on crime and vandalism and a lack of investment after apartheid. Those who can afford to, rely on generators, while those who can't are left in the dark.

Failure is also rampant in education. South Africa ranks 75 out of 76 on the league table of education systems developed by the OECD; only 37 percent of pupils starting school pass the matriculation exam and only 4 percent earn a degree; 27 percent of students who have attended school for 6 years still cannot read – compared with 4 percent in Tanzania and 19 percent in Zimbabwe; and after 5 years, only half can calculate 24 divided by 3 (The Economist 2017). These failings can be attributed to both the lack of teachers and teachers' lacking the knowledge and skills to perform their jobs.

It is also important to reflect on the history of transportation failures – indeed, transportation failure has been inherent to transportation planning in South Africa for quite a while. Perhaps, the most famous instance is Cape Town's Foreshore Freeway – an elevated highway built in 1977 as part of a modernist plan to expand the city into the bay – which was never completed. To this day, uncompleted elevated highway bridges hang disconnected from each other. The reason for this ostentatious failure has been the subject of much speculation – from engineering mishaps to government ineptitude to politicking. And, at different moments, the city has toyed with completing the road, even trying to include it in Cape Town's MyCiTi BRT. The extent to which the historical failures of this "freeway 'artefact'" (Kane 2011) impact transportation planning today depends on politics and power – like so much in South Africa.

There have also been times when potential failures were avoided – such as the case of personal rapid transit when it was debated in Johannesburg in 1977. The Zimbabwean Pneuways company proposed a design for a rubber-tire train running on an elevated concrete byway to operate between Johannesburg's city center and the neighborhoods of Braamfontein and Hillbrow. The system promised to move 30,000-seated passengers per hour at a cost R1 million per kilometer. It was eventually rejected by local transportation planners, who concluded that the monolithic system was inappropriate for the vast distances across low densities in South

African cities (The Star 1977). What remains to be seen is the extent to which transportation planners recognize and reflect upon these historical transportation failures in ongoing planning decisions.

These failings – both contemporary and historical, as well those in infrastructure, employment, and economic growth – came to the forefront on May 29, 2024 when South Africans voted in a general election that marked 30 years of enfranchisement for all. In this election, for the first time, the ANC failed to win the majority of seats in the National Parliament – winning just 40 percent of the vote, with the Democratic Alliance (DA) winning another 22 percent, uMkhonto we Sizwe (MK) winning 15 percent, and Economic Freedom Front (EFF) winning 9 percent of the vote (Chan 2024). The opposition parties – DA, MK, and EFF – recognized the ANC's decades of service delivery failures and promised, if elected, to fix these failures. These votes may provide an indication that the populace had lost its patience with failure. In the years to come, if the history of South African urban planning and development provides any indicator, it is that these other parties will find their own ways to fail.

It could be argued, however, that South Africa does not only need political change but also a shift in perspective. City leaders need to rethink what they perceive as success/failure, formal/informal, good/bad transportation – what constitutes a successful transportation system?; who would use and finance it? To what extent is informal transportation in South Africa failing?; what could constitute a formalized transportation system in South Africa? And, in what ways can informal transportation partner with formal transport? These are the types of complex provocations this book attempts to grapple with.

To move the dialogue forward, South African policymakers need to start searching for locally generated solutions rather than adopting global north solutions inappropriate to the South African context. My previous book uncovered the process whereby South African cities introduced BRT and how these decisions were influenced by city-to-city relations (Wood 2022a). The discussion in Chapter 4 of this book illustrates the manner by which economic, political, and social relations between and within cities were used by South African policymakers to advance (or subvert) particular actions and decisions regarding transit-oriented development (TOD). Instead, as discussions of decolonizing transportation in Chapter 2 of this book suggest, South African policymakers need to develop local solutions to local problems. This begins by strengthening relationships with local communities of (would-be) transportation users. Together, we can develop more effective solutions. Together, we can do more.

Reflecting on Transportation … Beyond

This book focuses on South African urban transportation but the analysis reaches far beyond. Indeed, I have not intended to craft a story unique to South Africa. Rather, this book provides a lens to rethink transportation in cities outside Africa and the global south. Similar transportation challenges span cities in the global north and south – poor air quality, low rates of electrification, insufficient public transportation infrastructure, increasing traffic congestion, and continued concerns for passenger and pedestrian safety are unfortunately commonplace in Accra, London, New York, and Sao Paolo. There are spectacular transportation failures like the Tacoma Narrows Bridge in Washington, which collapsed in 1940 just five months after completion, and the "Bridge to Nowhere" in northern California built in 1936 but never reconnected after the road washed away. And there are management failures like the near bankruptcy of London's Transport for London in 2022. In Africa and elsewhere, cities no longer operate public services instead relying on private and informal minibus services. Some measure this as state failure, while others interpret it as a measure of economic growth. Transportation failure is inherent to all cities around the world.

These, and similar transportation failures, are evident in Newcastle upon Tyne in the United Kingdom. According to a 2022 survey of the region by Sustrans, 49 percent of residents travel to work five or more days by walking, 36 percent by driving, 11 percent by public transportation, and 5 percent by cycling. This results in over 100 million car journeys of up to three miles in the region each year. If 80 percent of these trips moved to sustainable modes of transport, it could save 36,000 tons of greenhouse gas emissions. Of those surveyed, 65 percent would like to see more government spending on public transportation and nearly half argued for further spending on cycling or walking. The study also found, however, that there are differences in who is traveling: White people are more inclined to walk regularly than people of color (51 percent of White people versus 20 percent from ethnic minority groups), men cycle more frequently than women (22 percent of men versus 10 percent of women), and younger people feel safer cycling than older people (57 percent of people age 16–25 versus 38 percent of people aged 66+) (Sustrans 2022). These findings are especially hard-hitting as transportation is essential for economic revitalization, education, housing, and infrastructure investment in the postindustrial North East. In Newcastle, like in South Africa, transportation that fails to reach both urban and rural communities still holds the potential to stitch divided communities together.

These concerns came to the forefront in April 2024, when residents of the the region went to the polls to elect the region's first mayor. This first election for a regional mayor came after the seven local authorities agreed, as part of a national process of devolution, to create the North East Mayoral Combined Authority. The deal comes with £6 billion from the national government, of which £2 billion is dedicated to transportation infrastructure. Proposed projects include widening the A1 in Northumberland, reopening of the Leamside railway and the future of the Shields Ferry, as well as a plan for combining the competing bus services under public control through a franchising scheme (Richardson and Moss 2024). The importance of transportation within this broader political landscape was further noted at a hustings dedicated to transport. At this event, five candidates from five political parties debated their intended solutions: One advocated for greater investment in non-motorized modes like cycling and walking, as well as reducing the need to travel by investing in local communities; another argued for an expanded bus network, the "Angel Network" run by the local authority; and a third campaigned to make public transportation free for everyone under the age of 18 in addition to investing in active travel and a publicly-held bus network. To finance sustainable transport, a fourth committed to building Europe's largest windfarm. And, the fifth by contrast promised to improve the roads, reasoning that car-based travel was best for economic growth (Sustrans 2024). The plurality of responses exemplifies the challenge of solving transportation failings. And while there is no clear resolution, there is consensus that transportation is failing us everywhere.

In extending the analysis beyond South Africa, this book provides a lens for rethinking transportation within and beyond the sector. My intent has not been to vilify the exceptional efforts of hardworking planners and politicians, or shame those who are striving toward improvement; nor has it been to offer quick fixes to these failures by suggesting yet another set of ostensible successes from elsewhere. Rather, I have analyzed why transportation fails with the hope that one day it will be more convenient, sustainable, and inclusive for all.

Appendix A
Interview Schedule

List of Interviews taking place in South Africa in October 2019 and February 2020

Reference number	*Title*	*Organization*	*Place of interview*	*Date of interview*
1	Director[a]	Consultant	Johannesburg; conducted by telephone from Newcastle, UK	October 16, 2019
2	Lead Researcher	Consultant	Johannesburg; conducted by telephone from Newcastle, UK	October 18, 2019
3	Director	Consultant	Cape Town; conducted by telephone from Newcastle, UK	November 1, 2019
4	Project Manager	Consultant	Johannesburg	November 22, 2019
5	Professor	University of Witwatersrand	Johannesburg	November 25, 2019
6	Head of Planning, (2006–2010)[b]	Development Planning and Urban Management, City of Johannesburg	Johannesburg	November 25, 2019
7	Director	Consultant	Johannesburg	November 26, 2019
8	Owner	Transportation Operator	Johannesburg	November 26, 2019
9	MEC for Roads and Transport	Gauteng Province	Johannesburg	November 27, 2019
10	Head	Bus Operator	Johannesburg	November 27, 2019
11	Research Director	Gauteng City-Region Observatory	Johannesburg	November 27, 2019

(Continued)

(Continued)

List of Interviews taking place in South Africa in October 2019 and February 2020				
Reference number	*Title*	*Organization*	*Place of interview*	*Date of interview*
12	Transport specialist	Gauteng City-Region Observatory	Johannesburg	November 27, 2019
13	Environmental specialist	Gauteng City-Region Observatory	Johannesburg	November 27, 2019
14	Head of Strategic Planning (2001–2017)	City Transformation and Spatial Planning, City of Johannesburg	Nairobi, Kenya; conducted by telephone from Johannesburg	December 19, 2019
15	MMC for Roads and Transport (2006–2013)	City of Johannesburg	Johannesburg	January 21, 2020
16	Head	Gauteng Planning Division	Johannesburg	January 21, 2020
17	Head of Transport (2009–2019)	Department of Transport, City of Johannesburg	Johannesburg	January 22, 2020
18	Director	Transport Advocacy Organization	Johannesburg	January 22, 2020
19	Professor 1	University of Johannesburg	Johannesburg	January 22, 2020
20	Professor 2	University of Johannesburg	Johannesburg	January 22, 2020
21	MEC for Roads and Transport (2006–2013)	Gauteng Province	Johannesburg	January 23, 2020
22	Head of Legal and Compliance	Ekurhuleni Harambee project team	Ekurhuleni	January 23, 2020
23	Head of Operations and Industry Transition	Ekurhuleni Harambee project team	Ekurhuleni	January 23, 2020
24	Director: Infrastructure and Safety	Department of Transport, City of Johannesburg	Johannesburg	January 24, 2020
25	Operations Manager of the Rea Vaya	Department of Transport, City of Johannesburg	Johannesburg	January 24, 2020
26	Manager for Building Infrastructure	Department of Transport, City of Johannesburg	Johannesburg	January 24, 2020
27	Chief Executive Officer	Golden Arrow Bus Services	Cape Town	January 28, 2020
28	Director	Consultant		

List of Interviews taking place in South Africa in October 2019 and February 2020

Reference number	*Title*	*Organization*	*Place of interview*	*Date of interview*
29	Head of Transport (2006–2009)	Department of Transport, City of Johannesburg	Cape Town; conducted by telephone from Cape Town	January 29, 2020
30	Manager, MyCiTi	Department of Transport, City of Cape Town	Cape Town	January 29, 2020
31	Project Director	Consultant	Cape Town	January 29, 2020
32	Manager: Public Transport Infrastructure (1983–2016)	Department of Transport, City of Cape Town	Cape Town	January 30, 2020
33	Professor	University of Cape Town	Cape Town	January 30, 2020
34	Manager of Transport Planning	Department of Transport, City of Cape Town	Cape Town	January 30, 2020
35	Director	Consultant	Johannesburg	February 3, 2020
36	Head: Coverage Infrastructure Finance, Transport, Logistics and Water Sectors	Development Bank of Southern Africa	Johannesburg	February 4, 2020
37	Principal: Infrastructure Finance-Coverage Division, Transport, Logistics and Water	Development Bank of Southern Africa	Johannesburg	February 4, 2020
38	Head of Public Relations	Uber	Johannesburg	February 4, 2020
39	Executive Manager: Planning and Strategy	Johannesburg Development Agency, City of Johannesburg	Johannesburg	February 5, 2020
40	Professor	University of Pretoria	Tshwane	February 7, 2020[c]
41	Director: BRT Implementation	City of Johannesburg	Johannesburg	February 3, 2012
42	Director: Centre for Transport Studies	University of Cape Town	Cape Town	February 7, 2012

(*Continued*)

(*Continued*)

List of Interviews taking place in South Africa in October 2019 and February 2020				
Reference number	*Title*	*Organization*	*Place of interview*	*Date of interview*
43	MMC for Roads and Transport (2006–2013)	City of Johannesburg	Johannesburg	February 17, 2012
44	Executive Director	Victoria Transport Policy Institute	Vancouver, Canada; conducted by telephone from Johannesburg	February 20, 2012
45	Executive Director	Viva Cities	Manila, Phillipines; conducted by telephone from Johannesburg	February 23, 2012
46	Coordinator: Participatory Governance Program	PlanAct	Johannesburg	February 24, 2012
47	Chief Executive Officer	Johannesburg Development Agency	Johannesburg	March 6, 2012
48	Director	Luis Willumsen Consultancy	London, UK; conducted by telephone from Johannesburg	March 23, 2012
49	Researcher	University of Cape Town	Cape Town	April 4, 2012
50	Director	ITS Ltd	Cape Town	April 5, 2012
51	Mayoral Committee Member: Transport, Roads & Stormwater	City of Cape Town	Cape Town	April 10, 2012
52	Director	Pendulum Consultants	Cape Town	April 11, 2012
53	Director	Bicycle Empowerment Network	Cape Town	April 12, 2012
54	Director of Transport	City of Cape Town	Cape Town	April 13, 2012
55	Director: Engineering	HHO Africa	Cape Town	June 5, 2012

a Due to the sensitive nature of the interview material, the names of the interview respondents have been omitted. For those interviews that are both anonymous and confidential, a notation has been indicated; however, only a general title, the interview location, and the interview date have been included. No direct quotes have been used from those interviews.

b For those interview respondents who currently held other positions, I included the position that they were being discussed for.

c Fieldwork was curtailed by the outbreak of Covid-19 and subsequent travel restrictions as well as bereavement and two maternity leaves.

References

Adeoye, B. 2019. Colonial Transport System in Africa: Motives, Challenges and Impact. *African Journal of History and Archaeology* 4, no. 1: 14–26.

Adey, P., D. Bissell, K. Hannam, P. Merriman, and M. Sheller, eds. 2014. *Handbook of Mobilities*. London: Routledge.

Adorno, R. 1986. *Guaman Poma: Writing and Resistance in Colonial Peru*. Austin: University of Texas Press.

Agbiboa, D. 2020. How Informal Transport Systems Drive African Cities. *Current History* 119: 175–181.

Agbiboa, D. 2022. *They Eat Our Sweat: Transport Labor, Corruption, and Everyday Survival in Urban Nigeria*. Critical Frontiers of Theory, Research, and Policy in International Development Studies. Oxford, New York: Oxford University Press.

Agbiboa, D. 2023. Urban Taxi Slogans: The People's Arts. *African Arts* 56, no. 1: 42–61.

Agyemang, K.K., R.O. Amoako-Sakyi, K.B. Antwi, C.A. Mensah, and A.M. Abane. 2020. Transit Oriented Development: Theory and Implementation Challenges in Ghana. *TeMA - Journal of Land Use, Mobility and Environment* 13, no. 3 (December 31): 409–425.

Akbari, S., M.S. Mahmoud, A. Shalaby, and K.N. Habib. 2018. Empirical Models of Transit Demand with Walk Access/Egress for Planning Transit Oriented Developments around Commuter Rail Stations in the Greater Toronto and Hamilton Area. *Journal of Transport Geography* 68: 1–8.

Ake, C. 1979. *Social Science as Imperialism: The Theory of Political Development*. Ibadan University Press.

Alca˘ntara De Vasconcellos, E. 2004. The Use of Streets: A Reassessment and Tribute to Donald Appleyard. *Journal of Urban Design* 9, no. 1 (February): 3–22.

Alderman, D. 2018. The Racialized and Violent Biopolitics of Mobility in the USA: An Agenda for Tourism Geographies. *Tourism Geographies* 20, no. 4: 717–720.

Alderman, D., and J. Inwood. 2016. Mobility as Antiracism Work: The "Hard Driving" of NASCAR's Wendell Scott. *Annals of the American Association of Geographers* 106, no. 3: 597–611.

Alderman, D., P. Kingsbury, and O. Dwyer. 2013. Reexamining the Montgomery Bus Boycott: Toward an Empathetic Pedagogy of the Civil Rights Movement. *The Professional Geographer* 65, no. 1: 171–186.

Alderman, D., K. Williams, and E. Bottone. 2022. Jim Crow Journey Stories: African American Driving as Emotional Labor. *Tourism Geographies* 24, no. 2–3: 198–222.

Ali, S.H., and R. Keil. 2007. Contagious Cities. *Compass* 1, no. 5: 1207–1226.

Ali, S.H., and R. Keil. 2008. *Networked Disease: Emerging Infections in the Global City*. Oxford: Wiley-Blackwell.

Allara, P. 2017. Susan Woolf: Taxi Hand Signs: South African Jewish Museum Cape Town, South Africa July 10–September 5, 2016. *African Arts* 50, no. 4: 82–84.

Amundsen, C. 1995. Public Art and Public Transportation. In *Seventh National Conference on Light Rail Transit*, 75–86.

Anciaes, P.R., J. Nascimento, and S. Silva. 2017. The Distribution of Walkability in an African City: Praia, Cabo Verde. *Cities* 67: 9–20.

Anderson, B. 2019. Cultural Geography II: The Force of Representations. *Progress in Human Geography* 43, no. 6: 1120–1132.

Andreasen, M.H., and L. Møller-Jensen. 2017. Access to the City: Mobility Patterns, Transport and Accessibility in Peripheral Settlements of Dar Es Salaam. *Journal of Transport Geography* 62 (June): 20–29.

Appelhans, N., W. Scholz, and S. Baumgart, eds. 2021. *Transport Planning and Mobility in Urban East Africa*. Abingdon: Routledge.

Appleyard, D. 1981. *Liveable Streets*. Berkeley: University of California Press.

Arrive Alive. 2009. REA VAYA Expresses Delight with Artswork Programme. *Road Safety Highlights*, September 21. https://www.arrivealive.mobi/news.aspx?i=7667&s=3&page=REA-VAYA-expresses-delight-with-Artswork-Programme.

Atkins, P. 2012. *Animal Cities: Beastly Urban Histories*. Abingdon: Taylor & Francis.

Ayoubi, N. 2021. Meet Boji, Istanbul's Commuting Street Dog and Newest Influencer. *Middle East Eye*, October 6. https://www.middleeasteye.net/news/boji-dog-istanbul-commuting-influencer.

Baderoon, G. 2017. Animal Likenesses: Dogs and the Boundary of the Human in South Africa. *Journal of African Cultural Studies* 29, no. 3: 345–361.

Baker, L. 2020. Constructing Sub-Saharan African Mobilities through the Flow of Second-Hand Objects: Scripting Bicycles for Namibian Users. *Journal of Transport Geography* 83, no. 2: 1–7.

Baker, T., I.R. Cook, E. McCann, C. Temenos, and K. Ward. 2016. Policies on the Move: The Transatlantic Travels of Tax Increment Financing. *Annals of Association of American Geographers* 106, no. 2: 459–469.

Balme, C.B. 1999. *Decolonizing the Stage: Theatrical Syncretism and Post-Colonial Drama*. Oxford University Press.

Bantom, K. 2021. Taxi SA's First Mobile Art Gallery. *News24*, May 25, sec. News24. https://www.news24.com/news24/community-newspaper/peoples-post/taxi-sas-first-mobile-art-gallery-20210524.

Barton, H., S. Thompson, S. Burgess, and M. Grant. 2015. Introduction. In *The Routledge Handbook of Planning for Health and Well-Being: Shaping a Sustainable and Healthy Future*, ed. H. Barton, S. Thompson, S. Burgess, and M. Grant, xxxii–xxxiv. Oxford: Routledge.

Barton, H., and C. Tsourou, eds. 2000. *Healthy Urban Planning*. Oxford: Routledge.

Barua, M., and A. Sinha. 2019. Animating the Urban: An Ethological and Geographical Conversation. *Social & Cultural Geography* 20, no. 8: 1160–1180.

BBC News. 2020. Coronavirus: South Africa Eases Lockdown as "Outbreak Reaches Peak." *BBC News*, August 15, sec. Africa. https://www.bbc.com/news/world-africa-53795339.

Bear, C., K. Wilkinson, and L. Holloway. 2017. Visualizing Human-Animal-Technology Relations: Field Notes, Still Photography, and Digital Video on the Robotic Dairy Farm. *Society & Animals* 25: 225–256.

Beavon, K., and C. Rogerson. 1981. Trekking on: Recent Trends in the Human Geography of Southern Africa. *Progress in Human Geography* 5, no. 2: 159–189.

Beck, A.M. 1973. *The Ecology of Stray Dogs: A Study of Free-Ranging Urban Animals*. West Lafayette, Indiana: Purdue University Press.

de Beer, L. 2013. A Technical and Financial Assessment for the Viability of a Bike Sharing Scheme Resulting in a Sustainable Business Model and Contract Specification Johannesburg.

Beer, L., and D. Valjarevic. 2015. Bike Sharing in Johannesburg – Trendy Idea but Is It Financially Feasible? In *Proceedings of the 34th Southern African Transport Conference*, 499–518. Tshwane, South Africa.

Behrens, R. 2005. Accommodating Walking as a Travel Mode in South African Cities: Towards Improved Neighborhood Movement. *Planning Practice and Research* 20, no. 2: 163–182.

Behrens, R., and G. Makajuma. 2017. Pedestrian Crossing Behavior in Cape Town and Nairobi: Observations and Implications. In *Non-Motorized Transport Integration into Urban Transport Planning in Africa*, ed. W.V. Mitullah, M. Vanderschuren, and M. Khayesi, (pp. 27–56). London: Routledge.

Behrens, R., D. McCormick, and D. Mfinanga. 2015. *Paratransit in African Cities: Operations, Regulation and Reform*. New York: Earthscan.

Bell, D. 1978. *Race, Racism, and American Law*. New York: Aspen.

Belzer, D., G. Autler, J. Espinosa, S. Feigon, and G. Ohland. 2004. The Transit-Oriented Development Drama and Its Actors. In *The New Transit Town: Best Practices in Transit-Oriented Development*, ed. H. Dittmarr and G. Ohland, 41–56. Washington, DC: Island Press.

Benedictow, O.J. 2013. Yersinia Pestis, the Bacterium of Plague, Arose in East Asia. Did It Spread Westwards via the Silk Roads, the Chinese Maritime Expeditions of Zheng He or over the Vast Eurasian Populations of Sylvatic (Wild) Rodents? *Journal of Asian History* 47, no. 1: 1–31.

Bennett, C. 1960. Cultural Animal Geography: An Inviting Field of Research. *The Professional Geographer* 12, no. 5: 12–14.

Benton, J.S., G. Jennings, J. Walker, and J. Evans. 2023. "Walking Is Our Asset": How to Retain Walking as a Valued Mode of Transport in African Cities. *Cities* 137 (June): 104297.

van den Berg, C. 2007. *The Eland*. Johannesburg.

Best, A. 2016. The Way They Blow the Horn: Caribbean Dollar Cabs and Subaltern Mobilities. *Annals of the American Association of Geographers* 106, no. 2: 442–449.

Bickford, G., and R. Behrens. 2015. What Does Transit Oriented Development Mean in a South African Context? A Multiple Stakeholder Perspective from Johannesburg. In *Proceedings of the 34th Southern African Transport Conference*, 1–13. Tshwane, South Africa.

Bissell, D., and G. Fuller. 2017. Material Politics of Images: Visualising Future Transport Infrastructures. *Environment and Planning A: Economy and Space* 49, no. 11: 2477–2496.

Bledsoe, A. 2021. Methodological Reflections on Geographies of Blackness. *Progress in Human Geography* 45, no. 5: 1003–1021.

Bledsoe, A., L. Eaves, and B. Williams. 2017. Introduction: Black Geographies in and of the United States South. *Southeastern Geographer* 57, no. 1: 6–11.

Bledsoe, A., and W.J. Wright. 2019a. The Anti-Blackness of Global Capital. *Environment and Planning D: Society and Space* 37, no. 1: 8–26.

Bledsoe, A., and W.J. Wright. 2019b. The Pluralities of Black Geographies. *Antipode* 51, no. 2: 419–437.

Blench, R. 2004. The History and Spread of Donkeys in Africa. In *Donkeys, People and Development. A Resource Book of the Animal Traction Network for Eastern and Southern Africa*, ed. P. Starkey and D. Fielding, 9. Wageningen, The Netherlands: ACP-EU Technical Centre for Agricultural and Rural Cooperation (CTA).

Boonzaier, E., P. Berens, A. Smith, and C. Malherbe. 1996. *Cape Herders: A History of the Khoikhoi of southern Africa*, 54. Cape Town: David Philip Publishers.

Booysen, S., ed. 2016. *Fees Must Fall: Student Revolt, Decolonisation and Governance in South Africa*. Johannesburg, South Africa: Witwatersrand University Press.

Bownes, D., O. Green, and S. Mullins. 2012. *Underground: How the Tube Shaped London*. Allen Lane, London: Penguin Press.

Braczkowski, A.R., and D. Biggs. 2021. What COVID-19 Travel Bans Have Done to Conservation Tourism in Africa. *The Conversation*, October 19. https://theconversation.com/what-covid-19-travel-bans-have-done-to-conservation-tourism-in-africa-169324.

Brady, B. 2012. The Politics of the Pound: Controlling Loose Dogs in Nineteenth-Century New York City. *Jefferson Journal of Science and Culture* no. 2: 9–25.

Brand, A.L., K. Lowe, and E. Hall. 2020. Colorblind Transit Planning: Modern Streetcars in Washington, DC, and New Orleans. *Journal of Race, Ethnicity and the City* 1, no. 1–2: 87–108.

Brand, A.L., and C. Miller. 2020. Tomorrow I'll Be at the Table: Black Geographies and Urban Planning: A Review of the Literature. *Journal of Planning Literature* 35, no. 4: 460–474.

Brand, B. 1980. *Masstransiscope*. https://www.billbrand.net/public-art.

Breitbart, M., and P. Worden. 1994. Boston - Creating a Sense of Purpose: Public Art and Boston's Orange Line. *Places* 9, no. 2: 80–86.

Bremner, L. 2004. *Johannesburg: One City Colliding Worlds*. Johannesburg: Real African Publishers.

Brown, R.L. 2015. In Johannesburg, a New Bridge Must Also Address Apartheid's Legacy - Bloomberg. *Bloomberg City Lab*, February 2. https://www.bloomberg.com/news/articles/2015-02-02/in-johannesburg-a-new-bridge-must-also-address-apartheid-s-legacy.

Brown v. Board of Education of Topeka heard by the US Supreme Court in 1954.

347 in U.S. 483 (1954). https://supreme.justia.com/cases/federal/us/347/483/

Browder v. Gayle is 42 F. Supp. 707 (M.D. Ala. 1956), affirmed, 352 U.S. 903 (1956). https://law.justia.com/cases/federal/district-courts/FSupp/142/707/2263463/

Buchanan, C. 1963. *Traffic in Towns*. Harmondsworth: Penguin Books.

Bullard, R. 2004. The Anatomy of Transportation Racism. In *Highway Robbery: Transportation Racism & New Routes to Equity*, ed. R. Bullard, G. Johnson, and A. Torres, 15–31. Cambridge, MA: South End Press.

Buller, H. 2014. Animal Geographies I. *Progress in Human Geography* 38, no. 2: 308–318.

Buller, H. 2015. Animal Geographies II: Methods. *Progress in Human Geography* 39, no. 3: 374–384.

Buller, H. 2016. Animal Geographies III: Ethics. *Progress in Human Geography* 40, no. 3: 422–430.

Bunge, W. 2011. *Fitzgerald: Geography of a Revolution*. Athens: University of Georgia Press.

Bunn, D. 2008. Art Johannesburg and Its Objects. In *Johannesburg: The Elusive Metropolis*, ed. S. Nuttall and A. Mbembe, 137–169. London: Duke University Press.

Bunnell, T. 2015. Antecedent Cities and Inter-Referencing Effects: Learning from and Extending beyond Critiques of Neoliberalism. *Urban Studies* 52, no. 11: 1983–2000.

Business Tech. 2020. Lockdown Forced Nearly Half of Small Businesses in South Africa to Close: Study. *Business Tech*, December 7, sec. Business. https://businesstech.co.za/news/business/455100/lockdown-forced-nearly-half-of-small-businesses-in-south-africa-to-close-study/.

Butcher, B., and B. Milne. 2020. Coronavirus: How Will Transport Need to Change? *BBC News*, May 18, sec. Explainers. https://www.bbc.com/news/explainers-52534135.

Byrne, D. 2020. "Policing Borders": Extermination and Relocation of Insects in Three South African Texts. *Journal of Literary Studies* 36, no. 4: 86–104.

Cabaret, A. 2012. *Back to the Streets: Exploratory Research on Pedestrian Life and Walking Spaces in the Greater Johannesburg Area*. Johannesburg: University of the Witwatersrand.

Calthorpe, P. 1993. *The Next American Metropolis: Ecology, Community, and the American Dream*. Princeton, New Jersey: Princeton Architectural Press.

Campbell, M.O., and M.E. Torres Alvarado. 2011. Public Perceptions of Jaguars Panthera Onca, Pumas Puma Concolor and Coyotes Canis Latrans in El Salvador. *Area* 43, no. 3: 250–256.

Catsam, D.C. 2007. Marching in the "Dark City": Bus Boycotts in South Africa in the 1940s and the Limits and Promise of Comparative History. *Safundi* 8, no. 3: 315–325.

Cele, S., and K. Siwele. 2024. Housing Shortage in Africa's Richest City Leaves Squatters Risking Their Lives for Shelter. *Bloomberg.Com*, January 9. https://www.bloomberg.com/news/articles/2024-01-09/johannesburg-housing-shortage-leaves-hijacked-buildings-as-a-last-resort.

Cervero, R. 1993. *Transit-Supportive Development in the United States: Experiences and Prospects*. Washington, DC: Federal Transit Administration, U.S. Department of Transportation.

Cervero, R., and K. Kockelman. 1997. Travel Demand and the 3Ds: Density, Diversity, and Design. *Transportation Research Part D: Transport and Environment* 2, no. 3: 199–219.

Césaire, A. 1950. *Discours Sur Le Colonialisme*. New York, NY.

Chan, S. 2024. South Africa Election: ANC's Lost Majority Ushers in a New Era of Coalition Politics. *The Conversation*, June 3. https://theconversation.com/south-africa-election-ancs-lost-majority-ushers-in-a-new-era-of-coalition-politics-230927.

Chang, I.-C.C. 2017. Failure Matters: Reassembling Eco-Urbanism in a Globalizing China. *Environment and Planning A* 49, no. 8: 1719–1742.

Chatman, D.G., R. Xu, J. Park, and A. Spevack. 2019. Does Transit-Oriented Gentrification Increase Driving? *Journal of Planning Education and Research* 39, no. 4: 482–495.

Chen, N. 2019. Book Review: Beyond Mobility: Planning for Cities and Places. *Journal of Urban Affairs* 41, no. 8: 1227–1229.

Chinodya, S. 2017. *Can We Talk and Other Stories*. Bulawayo: Weaver Press.

Choi, K., and P. Denice. 2022. Inclusion, Walkability Will Be Key to Rebuilding Cities after the COVID-19 Pandemic. *The Conversation*, January 10. https://theconversation.com/inclusion-walkability-will-be-key-to-rebuilding-cities-after-the-covid-19-pandemic-174313.

Cirolia, L.R., and J. Harber. 2022. Urban Statecraft: The Governance of Transport Infrastructures in African Cities. *Urban Studies* 59, no. 12: 2431–2450.

City of Cape Town. 2003. *Provincial Vision for Public Transport and Five-Year Strategic Delivery Programme*. Cape Town: Provincial Administration of the Western Cape.

City of Cape Town. 2016. *Transit Oriented Development Strategic Framework*. Cape Town: Transport and Urban Development Authority.

City of Cape Town & Provincial Government of Western Cape. 2005. *Klipfontein Corridor Conceptual Urban Planning and Design and Non-Motorised Transport*. Cape Town.

City of Johannesburg. 2009. *Framework for Non-Motorized Transport*. Johannesburg: Department of Planning and Urban Management.

City of Johannesburg. 2011. *Johannesburg 2040: Growth and Development Strategy*. Johannesburg: Office of the City Manager.

City of Johannesburg. 2012a. *Strategic Area Framework: Empire Perth Development Corridor*. Johannesburg, South Africa.

City of Johannesburg. 2012b. *Strategic Area Framework: Louis Botha Development Corridor*. Johannesburg, South Africa.

City of Johannesburg. 2012c. *Strategic Area Framework: Turffontein Development Corridor*. Johannesburg, South Africa.

City of Johannesburg. 2013a. *Strategic Integrated Transport Framework*. Johannesburg: Department of Transport.

City of Johannesburg. 2013b. *Complete Streets Design Guidelines*. Johannesburg: City of Johannesburg.

City of Johannesburg. 2014a. Corridors of Freedom, a Vision of a City That Works for Its People. *Mail & Guardian*, June 26.

City of Johannesburg. 2014b. Making Johannesburg a Cycle Friendly City presented at the Public Consultation, October 17, Johannesburg.

City of Johannesburg. 2015. *Construction of the 'Great Walk' Bridge Commences*. Johannesburg: Johannesburg Development Agency.

Clarke, W.E. 1896. Bird Migration in the British Isles: Its Geographical and Meteorological Aspects. *Scottish Geographical Magazine* 12, no. 12: 616–626.

Classen, R., and J. Van der Post. 2020. Zola Budd, G-String, Dolphin: These Cars Have Strange Nicknames in South Africa. *News24*, June 16, sec. Life. https://www.news24.com/life/motoring/news/wildonwheels/zola-budd-g-string-dolphin-these-cars-have-strange-nicknames-in-south-africa-20200616.

Clayton, D.W. 2020. The Passing of "Geography's Empire" and Question of Geography in Decolonization, 1945–1980. *Annals of the American Association of Geographers* 110, no. 5: 1540–1558.

Clement, V. 2019. Beyond the Sham of the Emancipatory Enlightenment: Rethinking the Relationship of Indigenous Epistemologies, Knowledges, and Geography through Decolonizing Paths. *Progress in Human Geography* 43, no. 2: 276–294.

Cole, E. 1967. *House of Bondage*. 1st edition. New York: Aperture.

Connell, R. 2018. Decolonizing Sociology. *Contemporary Sociology* 47, no. 4: 399–407.

Connolly, C., R. Keil, and S.H. Ali. 2021. Extended Urbanisation and the Spatialities of Infectious Disease: Demographic Change, Infrastructure and Governance. *Urban Studies* 58, no. 2 (February 1): 245–263.

Cooke, S., R. Behrens, and M. Zuidgeest. 2018. The Relationship between Transit-Oriented Development, Accessibility and Public Transport Viability in South African Cities: A Literature Review and Problem Framing. In *Proceedings of the 37th Southern African Transport Conference*, 365–380. Tshwane, South Africa.

Cooke, S., B. Ryseck, G. Siame, A. Nkurunziza, L. Molefe, and M. Zuidgeest. 2022. Proximity Is Not Access: A Capabilities Approach to Understanding Non-Motorized Transport Vulnerability in African Cities. *Frontiers in Sustainable Cities* 4: 1–13.

Correal, A. 2018. Inside the Dollar Van Wars. *The New York Times*, June 8, sec. New York. https://www.nytimes.com/2018/06/08/nyregion/inside-the-dollar-van-wars.html.

Correia, S., S. Luck, and E. Verner. 2020. Pandemics Depress the Economy, Public Health Interventions Do Not: Evidence from the 1918 Flu. *SSRN Electronic Journal* 82, no. 4: 917–957.

COWI. 2012. *Pre-Feasibility Study: Cycling Programme in the City of Johannesburg*. Denmark: COWI.

Craggs, R. 2014. Postcolonial Geographies, Decolonization, and the Performance of Geopolitics at Commonwealth Conferences. *Singapore Journal of Tropical Geography* 35, no. 1: 39–55.

Craggs, R., and H. Neate. 2016. Post-Colonial Careering and Urban Policy Mobility: Between Britain and Nigeria, 1945–1990. *Transactions of the Institute of British Geographers*.42, no. 1: 44–57.

Craggs, R., and H. Neate. 2019. What Happens If We Start from Nigeria? Diversifying Histories of Geography. *Annals of the American Association of Geographers*: 110, no. 3: 899–916.

Craggs, R., and H. Neate. 2023. *Decolonising Geography? Disciplinary Histories and the End of the British Empire in Africa, 1948–1998*. London: Wiley. https://www.wiley.com/en-gb/Decolonising+Geography%3F+Disciplinary+Histories+and+the+End+of+the+British+Empire+in+Africa%2C+1948+1998-p-9781119549307.

Crenshaw, K. 1991. Mapping the Margins: Intersectionality, Identity Politics, and Violence against Women of Color. *Stanford Law Review* 43, no. 6: 1241.

Crenshaw, K. 2010. Twenty Years of Critical Race Theory: Looking Back to Move Forward. *Connecticut Law Review* 43: 1253.

Crenshaw, K., N. Gotanda, G. Peller, and K. Thomas. 1995. *Critical Race Theory: The Key Writings That Formed the Movement*. New York: New Press.

Cresswell, T. 2010. Towards a Politics of Mobility. *Environment and Planning D: Society and Space* 28: 17–31.

Cresswell, T. 2014. Mobilities III: Moving On. *Progress in Human Geography* 38, no. 5: 712–721.

Croese, S., and A. Wood. 2024. African Urban Studies: Contributions and Challenges. *Urban Studies* 61, no. 13: 2477–2491.

Crot, L. 2010. Transnational Urban Policies: "Relocating" Spanish and Brazilian Models of Urban Planning in Buenos Aires. *Urban Research & Practice* 3, no. 2: 119–137.

Crush, J. 1993. The Discomforts of Distance: Post-Colonialism and South African Geography. *South African Geographical Journal* 75, no. 2: 60–68.

Daigle, M., and J. Sundberg. 2017. From Where We Stand: Unsettling Geographical Knowledges in the Classroom. *Transactions of the Institute of British Geographers* 42, no. 3: 338–341.

Daley, P.O., and A. Murrey. 2022. Defiant Scholarship: Dismantling Coloniality in Contemporary African Geographies. *Singapore Journal of Tropical Geography* 43, no. 2: 159–176.

Debrie, J. 2010. From Colonization to National Territories in Continental West Africa: Thehistorical Geography of a Transport Infrastructure Network. *Journal of Transport Geography* 18: 292–300.

Department of Community Development. 2006. *Public Art Policy*. City of Johannesburg.

Desai, V. 2017. Black and Minority Ethnic (BME) Student and Staff in Contemporary British Geography. *Area* 49, no. 3: 320–323.

Deskins, D. 1969. Geographical Literature on the American Negro, 1949-1968: A Bibliography. *The Professional Geographer* 21: 145–149. World.

Deskins, D. 1972. Race, Residence, and Workplace in Detroit, 1880 to 1965. *Economic Geography* 48, no. 1: 79–94.

Didier, S., E. Peyroux, and M. Morange. 2012. The Spreading of the City Improvement District Model in Johannesburg and Cape Town: Urban Regeneration and the Neoliberal Agenda in South Africa. *International Journal of Urban and Regional Research* 36, no. 5: 915–935.

Diemen, E. van. 2021. CORONAVIRUS VARIANT: Omicron: Domino Effect as World Puts Southern Africa in Lockdown – And New Cases Are Detected Abroad. *Daily Maverick*, November 28. https://www.dailymaverick.co.za/article/2021-11-28-omicron-domino-effect-as-world-puts-southern-africa-in-lockdown-and-new-cases-are-detected-abroad/.

Dillette, A., and S. Benjamin. 2022. The Black Travel Movement: A Catalyst for Social Change. *Journal of Travel Research* 61, no. 3: 463–476.
Diouf, I., F. Arroyo, W. Jia, C. Ochoa, and F. Taillandier. 2020. *Urban Mobility and Covid-19 in Africa*. Washington, DC: World Bank.
Dittmar, H., D. Belzer, and G. Autler. 2004. An Introduction to Transit-Oriented Development. In *The New Transit Town: Best Practices in Transit-Oriented Development*, ed. H. Dittmarr and G. Ohland, 1–18. Washington, DC: Island Press.
Dlamini, P. 2020. More Joburg Taxi Associations Plan to Hike Fares. *SowetanLIVE*, June 9. https://www.sowetanlive.co.za/news/south-africa/2020-06-09-more-joburg-taxi-associations-plan-to-hike-fares/.
Doble, J. 2020. Can Dogs Be Racist? The Colonial Legacies of Racialized Dogs in Kenya and Zambia. *History Workshop Journal* 89: 68–89.
Doherty, J. 2022. Motorcycle Taxis, Personhood, and the Moral Landscape of Mobility. *Geoforum* 136 (November): 242–250.
Donkin, R.A. 1985. *The Peccary: With Observations on the Introduction of Pigs to the New World.* Philadelphia, PA: American Philosophical Society.
Dovey, K., and E. Pafka. 2020. What Is Walkability? The Urban DMA. *Urban Studies* 57, no. 1: 93–108.
Drake, St.C., and H.R. Cayton. 1945. *Black Metropolis: A Study of Negro Life in a Northern City.* Chicago, IL: University of Chicago Press.
Driskill, Q.L., C. Finley, B.J. Gilley, and S.L. Morgensen. 2011. *Queer Indigenous Studies: Critical Interventions in Theory, Politics, and Literature.* Tuscon: University of Arizona Press.
Driver, F. 1992. Geography's Empire: Histories of Geographical Knowledge. *Environment and Planning D: Society and Space* 10, no. 1: 23–40.
Du Bois, W.E.B. 1899. *The Philadelphia Negro: A Social Study*. Philadelphia: University of Pennsylvania Press.
Duffy, R., and L. Moore. 2010. Neoliberalising Nature? Elephant-Back Tourism in Thailand and Botswana. *Antipode* 42, no. 3: 742–766.
Dugard, J. 2001. From Low Intensity War to Mafia War: Taxi Violence in South Africa (1987-2000). *Violence and Transition* 4: 1–46.
Dwyer, O.J. 1997. Geographical Research about African Americans: A Survey of Journals, 1911–1995. *The Professional Geographer* 49, no. 4: 441–451.
Eaves, L. 2020. Fear of an Other Geography. *Dialogues in Human Geography* 10, no. 1: 34–36.
Elden, S. 2009. Reassessing Kant's Geography. *Journal of Historical Geography* 35, no. 1: 3–25.
Elliot-Cooper, A. 2017. "Free, Decolonised Education": A Lesson from the South African Student Struggle. *Area* 49, no. 3: 332–334.
Emmett, M. 2021. Reigniting William Kentridge and Gerhard Marx's "The Firewalker." *Daily Maverick*, July 14. https://www.dailymaverick.co.za/article/2021-07-14-reigniting-william-kentridge-and-gerhard-marxs-the-firewalker/.
Enright, T. 2023. Art in Transit: Mobility, Aesthetics and Urban Development. *Urban Studies* 60, no. 1: 67–84.
Enright, T., and K. Ward. 2021. Governing Urban Infrastructures under Pandemic Conditions: Some Thoughts. *Urban Geography* 42, no. 7: 1023–1032.
Espi, L. 2017. *The Art of MyCiTi*. Cape Town: Transport and Urban Development Authority.
Esson, J. 2018. "The Why and the White": Racism and Curriculum Reform in British Geography Area 52: 708–715.

Esson, J., K.V. Gough, D. Simon, E.F. Amankwaa, O. Ninot, and P. Yankson. 2016. Livelihoods in Motion: Linking Transport, Mobility and Income-Generating Activities. *Journal of Transport Geography* 5: 182–188.

Esson, J., P. Noxolo, R. Baxter, P. Daley, and M. Byron. 2017. The 2017 RGS-IBG Chair's Theme: Decolonising Geographical Knowledges, or Reproducing Coloniality? *Area* 49, no. 3: 384–388.

Evans, J., J. O'Brien, and B.C. Ng. 2018. Towards a Geography of Informal Transport: Mobility, Infrastructure and Urban Sustainability from the Back of a Motorbike. *Transactions of the Institute of British Geographers* 34, no. 4: 674–688.

Ewing, R., and K. Bartholomew. 2013. *Pedestrian & Transit-Oriented Design*. Washington, DC: Urban Land Institute.

Ewing, R., and R. Cervero. 2010. Travel and the Built Environment: A Meta-Analysis. *Journal of the American Planning Association* 76, no. 3: 265–294.

Faiver-Serna, C. 2019. Juntxs/Together: Building Latinx Geographies. Society and Space.

Fanon, F. 1961. *The Wretched of the Earth*. London: Penguin Press.

Farmer, B. 2023. How Rolling Power Cuts Have Brought Cape Town – And South Africa – To Its Knees. *The Telegraph*, April 22. https://www.telegraph.co.uk/global-health/terror-and-security/south-africa-power-cuts-blackouts/.

Feldman, M.B. 2009. Where the Wild Things Aren't: Animals in New York City. *Minnesota Review* 2010, no. 73–74: 231–242.

Ferretti, F., and B. Viotto Pedrosa. 2018. Inventing Critical Development: A Brazilian Geographer and His Northern Networks. *Transactions of the Institute of British Geographers* 43, no. 4: 703–717.

Finden, S. 2011. *Casper the Commuting Cat: The True Story of the Cat Who Rode the Bus and Stole Our Hearts*. London, UK: Simon & Schuster UK.

Fitzpatrick, T. 2009. *Art and the Subway: New York Underground*. New Brunswick, NJ: Rutgers University Press.

Flack, A.J.P. 2016. Lions Loose on a Gentleman's Lawn: Animality, Authenticity and Automobility in the Emergence of the English Safari Park. *Journal of Historical Geography* 54: 38–49.

Fletcher, T., and L. Platt. 2018. (Just) a Walk with the Dog? Animal Geographies and Negotiating Walking Spaces. *Social & Cultural Geography* 19, no. 2: 211–229.

Florida, R., A. Rodríguez-Pose, and M. Storper. 2021. Cities in a Post-COVID World. *Urban Studies* (June 27). https://doi.org/10.1177/00420980211018072.

Fobosi, D.S.C. 2020. South Africa's Minibus Taxi Industry Has Been Marginalised for Too Long. This Must Change. *The Conversation*, July 14. https://theconversation.com/south-africas-minibus-taxi-industry-has-been-marginalised-for-too-long-this-must-change-142060.

Forsyth, A. 2015. What Is a Walkable Place? The Walkability Debate in Urban Design. *URBAN DESIGN International* 20, no. 4: 274–292.

Forsyth, A., and M. Southworth. 2008. Cities Afoot—Pedestrians, Walkability and Urban Design. *Journal of Urban Design* 13, no. 1: 1–3.

Gandy, M. 2004. Rethinking Urban Metabolism: Water, Space and the Modern City. *City* 8, no. 3: 363–379.

Gauteng Department of Roads and Transport. 2020. *Gauteng Province Household Travel Survey Report*. Tshwane, South Africa: Gauteng Department of Roads and Transport.

Gauvin, P. 1994. Peter Calthorpe: The Way Cities Oughta Be. *Palo Alto Online*, August 17. https://www.paloaltoonline.com/weekly/morgue/monthly/1994_Aug_17.CALTHORP.html.

Gaylard, G. 2017. An African Flaneur? Walking in Johannesburg in Portrait with Keys. In *Cities in Flux: Metropolitan Spaces in South African Literary and Visual Texts*, ed. O. Moreillon, A. Muller, and L. Stiebel, 57–72. Zurich: Lit Verlag.

Gevisser, M. 2014. *Lost and Found in Johannesburg*. Johannesburg: Jonathan Ball.

Gibbs, L. 2020. Animal Geographies I: Hearing the Cry and Extending Beyond. *Progress in Human Geography* 44, no. 4: 769–777.

Gibbs, L. 2021. Animal Geographies II: Killing and Caring (in Times of Crisis). *Progress in Human Geography* 45, no. 2: 371–381.

Gill, F. 1961. *Cape Trams: From Horse to Diesel*. Cape Town: Cape Electric Tramways.

Gill, J. 2018. Decolonizing Literature and Science. *Configurations* 26: 283–288.

Gillespie, K., and R.-C. Collard, eds. 2015. *Critical Animal Geographies: Politics, Intersections, and Hierarchies in a Multispecies World*. 1st ed. London, UK: Routledge.

Gilmore, R.W. 2007. *Golden Gulag: Prisons, Surplus, Crisis, and Opposition in Globalizing California*. Berkeley: University of California Press.

Glaeser, E. 2020. Cities and Pandemics Have a Long History. *City Journal*. https://www.city-journal.org/cities-and-pandemics-have-long-history.

Glaeser, E., C. Gorback, and S. Redding. 2020. *How Much Does COVID-19 Increase with Mobility? Evidence from New York and Four Other U.S. Cities*. Cambridge, MA: National Bureau of Economic Research.

Goldbaum, C. 2020. Subway Service Is Cut by a Quarter Because of Coronavirus. *New York Times*, March 24.

Goldsmith, S. 1963. *Designing for the Disabled: The New Paradigm*. London: Routledge.

Gomi, T. 1988. Bus Stops. Chronicle Books.

González, S. 2011. Bilbao and Barcelona "In Motion": How Urban Regeneration "Models" Travel and Mutate in the Global Flows of Policy Tourism. *Urban Studies* 48, no. 7: 1397–1418.

Goodfellow, T., and Z. Huang. 2021. Contingent Infrastructure and the Dilution of "Chineseness": Reframing Roads and Rail in Kampala and Addis Ababa. *Environment and Planning A: Economy and Space* 53, no. 4: 655–674.

Gounden, T. 2023. Rea Vaya Has No Answers for Thursday's Temporary Suspension of Services. *Eyewitness News*, March 23. https://ewn.co.za/0001/01/01/rea-vaya-has-no-answers-for-thursday-s-temporary-suspension-of-services. Accessed on February 1, 2024.

Grady, D. 2012. Thelma Glas, Who Helped Organize Alabama Bus Boycott, Dies at 96. *The New York Times*, July 27.

Grant, C., and D. Price. 2020. Decolonizing Art History. *Art History* 43, no. 1: 8–66.

Gray, M., J. Coates, M.Y. Bird, and T. Hetherington. 2016. *Decolonizing Social Work*. London, UK: Routledge.

Green-Simms, L.B. 2017. *Postcolonial Automobility: Car Culture in West Africa*. Minneapolis, Minnesota: University of Minnesota Press.

Gromley, A. 1998. *Angel of the North*. Gateshead, UK.

Gruzinski, S. 2002. *The Mestizo Mind: The Intellectual Dynamics of Colonization and Globalization*. London, UK: Routledge.

Guerra, E., R. Cervero, and S. Al. 2017. *Beyond Mobility: Planning Cities for People and Places*. London: Island Press.

Guinard, P. 2018. Redefining Publics, Artists, and Urban Spaces: The Case of Made in Musina, South Africa. *City & Society* 30, no. 1: 117–135.

Guinard, P., and A. Margier. 2018. Art as a New Urban Norm: Between Normalization of the City through Art and Normalization of Art through the City in Montreal and Johannesburg. *Cities* 77: 13–20.

Guinard, P., and G. Molina. 2018. Urban Geography of Arts: The Co-Production of Arts and Cities. *Cities* 77: 1–3.

Gurney, K. 2018. Zombie Monument: Public Art and Performing the Present. *Cities* 77: 33–38.

Habib, A. 2019. *Rebels and Rage: Reflecting on #FeesMustFall.* London: Jonathan Ball.

Haffajee, F. 2020. Covid-19's Impact: The Day the Bottom Fell out of South Africa – A Triple Pandemic Has Hit Us. *Daily Maverick*, July 15. https://www.dailymaverick.co.za/article/2020-07-15-the-day-the-bottom-fell-out-of-south-africa-a-triple-pandemic-has-hit-us/.

Hall, T. 2007. Artful Cities: Artful Cities. *Geography Compass* 1, no. 6: 1376–1392.

Hamidi, S., S. Sabouri, and R. Ewing. 2020. Does Density Aggravate the COVID-19 Pandemic? *Journal of the American Planning Association* 86, no. 4: 495–509.

Hanekom, D. 2004. The Use of Donkeys for Transport in South Africa. In *Donkeys, People and Development. A Resource Book of the Animal Traction Network for Eastern and Southern Africa*, ed. P. Starkey and D. Fielding. Wageningen, The Netherlands: ACP-EU Technical Centre for Agricultural and Rural Cooperation, (pp. 192–195).

Hannam, K., M. Sheller, and J. Urry. 2006. Mobilities, Immobilities and Moorings. *Mobilities* 1, no. 1: 1–22.

Hansen, T.B. 2006. Sounds of Freedom: Music, Taxis, and Racial Imagination in Urban South Africa. *Public Culture* 18, no. 1: 185–208.

Hansen, T.B. 2012. *Melancholia of Freedom: Social Life in an Indian Township in South Africa.* Princeton, NJ: Princeton University Press.

Hardy, D. 1992. The Garden City Campaign: An Overview. In *The Garden City: Past, Present and Future*, ed. S.V. Ward, 187–209. London: E & FN Spon.

Harradine, M. 1997. *Port Elizabeth: A Social Chronicle to the End of 1945.* Port Elizabeth: E.H. Walton Packaging.

Harris, A., and S. Moore. 2013. Planning Histories and Practices of Circulating Urban Knowledge. *International Journal of Urban and Regional Research* 37, no. 5: 1499–1509.

Harris, J. 2020. *The Subway Seeded the Massive Coronavirus Epidemic in New York City.* Cambridge, MA: National Bureau of Economic Research.

Harrison, K. 2014. *Aesthetic, Evocative and Invisible?* Johannesburg, South Africa: School of Architecture and Planning, University of the Witwatersrand and the Johannesburg Development Agency.

Harrison, P. 2015. South–South Relationships and the Transfer of "Best Practice": The Case of Johannesburg, South Africa. *International Development Planning Review* 37, no. 2: 205–223.

Harrison, P. 2020. *Johannesburg and Its Epidemics: Can We Learn from History?* Johannesburg: Gauteng City-Region Observatory. https://www.gcro.ac.za/outputs/occasional-papers/detail/johannesburg-and-its-epidemics-can-we-learn-from-history/.

Harrison, P., M. Rubin, A. Appelbaum, and R. Dittgen. 2019. Corridors of Freedom: Analyzing Johannesburg's Ambitious Inclusionary Transit-Oriented Development. *Journal of Planning Education and Research* 39, no. 4: 456–468.

Hart, G.H.T. 1984. Urban Transport, Urban Form and Discrimination in Johannesburg. *South African Geographical Journal* 66, no. 2: 152–167.

Hart, J. 2016. *Ghana on the Go: African Mobility in the Age of Motor Transportation.* Bloomington: Indiana University Press.

Harvey, D. 1972. Revolutionary and Counter-Revolutionary Theory in Geography and the Problem of Ghetto Formation. *Antipode* 4, no. 2: 1–13.

Harvey, D. 1989. *The Condition of Postmodernity.* Oxford: Blackwell Publishers Ltd.

Hawkins, H. 2013. Geography and Art: An Expanding Field: Site, the Body and Practice. *Progress in Human Geography* 37, no. 1: 52–71.

Hawkins, H. 2019. Geography's Creative (Re)Turn: Toward a Critical Framework. *Progress in Human Geography* 43, no. 6: 963–984.

Hawthorne, C. 2019. Black Matters Are Spatial Matters: Black Geographies for the Twenty-First Century. *Geography Compass* 13, no. 11: 1–13. https://onlinelibrary.wiley.com/doi/abs/10.1111/gec3.12468.

Hinger, B. 2022. Toward an Expanded Approach on Black Mobilities. *Progress in Human Geography* 46, no. 2: 359–371.

Hiralal, K., and Z. Jinnah, eds. 2018. *Gender and Mobility in Africa: Borders, Bodies and Boundaries*. London: Springer International Publishing. https://link.springer.com/10.1007/978-3-319-65783-7.

Hlalethwa, Z. 2018. The Art of Taxi Art. *The Mail & Guardian*, May 11. https://mg.co.za/article/2018-05-11-00-the-art-of-taxi-art/.

Ho, U. 2018. Circle of City Life Returns to This New/Old Part of Joburg. *Daily Maverick*. https://www.dailymaverick.co.za/article/2018-07-11-circle-of-city-life-returns-to-this-new-old-part-of-joburg/.

Hodgetts, T., and J. Lorimer. 2020. Animals' Mobilities. *Progress in Human Geography* 44, no. 1: 4–26.

Holmberg, T. 2015. *Urban Animals: Crowding in Zoocities*. London: Routledge.

Honey-Rosés, J., I. Anguelovski, V.K. Chireh, C. Daher, C. Konijnendijk van den Bosch, J.S. Litt, V. Mawani et al. 2020. The Impact of COVID-19 on Public Space: An Early Review of the Emerging Questions – Design, Perceptions and Inequities. *Cities & Health* 5, no. S1 (July 31): , S263–S279.

Hook, W., and A. Weinstock. 2021. The Role of BRT in Post-Pandemic South Africa. *Medium*, May 5. https://reorientations.medium.com/the-role-of-brt-in-post-pandemic-south-africa-3f3d8aff10f0.

Hovorka, A. 2008. Transspecies Urban Theory: Chickens in an African City. *Cultural Geographies* 15, no. 1: 95–117.

Hovorka, A. 2017. Animal Geographies I: Globalizing and Decolonizing. *Progress in Human Geography* 41, no. 3: 382–394.

Hovorka, A. 2018. Animal Geographies II: Hybridizing. *Progress in Human Geography* 42, no. 3: 453–462.

Hovorka, A. 2019. Animal Geographies III: Species Relations of Power. *Progress in Human Geography* 43, no. 4: 749–757.

Hovorka, A., S. McCubbin, and L. Van Patter. 2021. *A Research Agenda for Animal Geographies*. Cheltenham, UK or Northampton, Massachusetts: Edward Elgar Publishing. https://www.elgaronline.com/view/edcoll/9781788979986/9781788979986.xml.

Huang, J., and M.-P. Kwan. 2021. Uncertainties in the Assessment of COVID-19 Risk: A Study of People's Exposure to High-Risk Environments Using Individual-Level Activity Data. *Annals of the American Association of Geographers*: 968–987.

Hubbard, P., and A. Brooks. 2021. Animals and Urban Gentrification: Displacement and Injustice in the Trans-Species City. *Progress in Human Geography* 45, no. 6: 1490–1511.

Hubbard, S.M.L., M. Smalley, and S. Hubbard. 2010. Art in Transportation: Multimodal Overview. *Transportation Research Record: Journal of the Transportation Research Board* 2144, no. 1: 168–180.

Human, L., and E. Morrison. 2023. How the Taxi Industry Works in the Western Cape. *GroundUp News*, August 10. https://groundup.org.za/article/how-does-taxi-industry-work-in-western-cape/.

Institute for Transportation and Development Policy. 2021. Transit-Oriented Development Case Studies Report: An Assessment of Six Settlements and Interventions in East Africa and India.

Inwood, J., D. Alderman, and J. Williams. 2015. "Where Do We Go from Here?": Transportation Justice and the Struggle for Equal Access. *Southeastern Geographer* 55, no. 4: 417–433.

Isnard, H. 1971. *Géographie de La Décolonisation*. Paris: PUF.

ITDP. 2017. *The TOD Standard 3.0*. New York: Institute for Transportation and Development Policy.

ITDP. 2018. *Pedestrians First: A Tool for Walkable Cities*. New York City: Institute for Transportation and Development Policy.

Jackson, A. 2003. *Public Transport in Durban - A Brief History*. https://www.fad.co.za/Resources/transport/transport.htm.

Jacobs, J. 1961. *The Death and Life of Great American Cities*. New York: Vintage Books.

Jacobs, N.J. 2001. The Great Bophuthatswana Donkey Massacre: Discourse on the Ass and the Politics of Class and Grass. *The American Historical Review* 106, no. 2: 485–507.

Jacobsen, M. 2023. *Assembling Bus Rapid Transit in the Global South: Translating Global Models, Materialising Infrastructure Politics*. London: Routledge.

Jamme, H.-T., J. Rodriguez, D. Bahl, and T. Banerjee. 2019. A Twenty-Five-Year Biography of the TOD Concept: From Design to Policy, Planning, and Implementation. *Journal of Planning Education and Research* 39, no. 4: 409–428.

Jazeel, T. 2017. Mainstreaming Geography's Decolonial Imperative. *Transactions of the Institute of British Geographers* 42, no. 3: 334–337.

JDA. 2015a. *Great Walk Will Link Alex and Sandton over the M1*. Johannesburg: Johannesburg Development Agency.

JDA. 2015b. *Construction of R130m Alex-Sandton Bridge Gets under Way*. Johannesburg: Johannesburg Development Agency.

JDA. 2016. *Westbury's New Corridor of Freedom Landmark Starts Taking Shape*. Johannesburg: Johannesburg Development Agency.

JDA. 2018a. *Westdene Dam Precinct Interventions/NMT*. Johannesburg: Johannesburg Development Agency.

JDA. 2018b. *Non-Motorised Transport (NMT) Project Under Construction - Westbury Development Renewal Precinct – NMT Phase 4*. Johannesburg: Johannesburg Development Agency.

JDA. 2020. *Milpark Pedestrian Bridge Reaches Practical Completion*. Johannesburg: Johannesburg Development Agency.

Jennings, G. 2015a. Finding Our Balance: Considering the Opportunities for Public Bicycle Systems in Cape Town, South Africa. *Research in Transportation Business & Management* 15: 6–14.

Jennings, G. 2015b. Public Transport Interventions and Transport Justice in South Africa: A Literature and Policy Review. In *Proceedings of the 34th Southern African Transport Conference*. Tshwane, South Africa.

Jerolmack, C. 2008. How Pigeons Became Rats: The Cultural-Spatial Logic of Problem Animals. *Social Problems* 55, no. 1: 72–94.

Joubert, B. 1995. An Historical Perspective on Animal Power Use in South Africa. In *Animal Traction in South Africa: Empowering Rural Communities*, ed. P. Starkey, Wageningen, The Netherlands: ACP-EU Technical Centre for Agricultural and Rural Cooperation 125–153.

Joyce, P., ed. 1981. *South Africa's Yesterdays*. Johannesburg: Reader's Digest Association of South Africa.

Ka'bange, A., D. Mfinanga, and E. Hema. 2014. Paradoxes of Establishing Mass Rapid Transit Systems in African Cities – A Case of Dar Es Salaam Rapid Transit (DART) System, Tanzania. *Research in Transportation Economics* 48: 176–183.

Kane, L. 2011. Building the Foreshore Freeways: The Politics of a Freeway "Artefact." In *Proceedings of the 30th Southern African Transport Conference*, 131–139. Tshwane, South Africa.

Kar, A., H.T.K. Le, and H.J. Miller. 2021. What Is Essential Travel? Socioeconomic Differences in Travel Demand in Columbus, Ohio, during the COVID-19 Lockdown. *Annals of the American Association of Geographers* 112, no. 4: 1023–1046.

Kaur, T., S. Forde, and L. Hayhurst. 2023. Cycling Through the Buffer Zone: Working-Class Cyclists on Rudolf Greyling Street, Bloemfontein. *Urban Forum* 34, no. 4 (December): 441–462.

Kekana, M., and S. Allison. 2019. Bridges Only Work When Both Sides Cross Them. *YES! Magazine*, November 12, sec. Social Justice.

Kelaidis, K. 2020. What the Great Plague of Athens Can Teach Us Now. *The Atlantic*. https://www.theatlantic.com/ideas/archive/2020/03/great-plague-athens-has-eerie-parallels-today/608545/.

Kelson, C., and N. Kotze. 2016. From the Randse Afrikaanse Universiteit to the University of Johannesburg: An Inclusive South African University. In *The Origin and Growth of Geography as a Discipline at South Africa Universities*, ed. G. Visser, R. Donaldson, and C. Seethal, 327–348. Stellenbosch: Sun Press.

Kendall, C., P. Hudelson, E. Leontsini, P. Winch, L. Lloyd, and F. Cruz. 1991. Urbanization, Dengue, and the Health Transition: Anthropological Contributions to International Health. *Medical Anthropology Quarterly* 5, no. 3: 257–268.

Kentridge, W., and G. Marx. 2009. *Firewalker*. Johannesburg.

Kessi, S., Z. Marks, and E. Ramugondo. 2020. Decolonizing African Studies. *Critical African Studies* 12, no. 3: 271–282.

Khayesi, M., H. Monheim, and J.M. Nebe. 2010. Negotiating "Streets for All" in Urban Transport Planning: The Case for Pedestrians, Cyclists and Street Vendors in Nairobi, Kenya. *Antipode* 42, no. 1: 103–126.

Khosa, M. 1990. Changing Patterns of "Black" Bus Subsidies in the Apartheid City, 1944-1986. *GeoJournal* 22, no. 3: 251–259.

Khosa, M. 1991. Routes, Ranks and Rebels: Feuding in the Taxi Revolution. *Journal of Southern African Studies* 18, no. 1: 232–251.

Khosa, M. 1992. Routes, Ranks and Rebels: Feuding in the Taxi Revolution. *Journal of Southern African Studies* 18, no. 1: 232–251.

Khosa, M. 1995. Transport and Popular Struggles in South Africa. *Antipode* 27, no. 2: 167–188.

Kleinman, M. 2020. Policy Challenges for the Post-Pandemic City. *Environment and Planning B: Urban Analytics and City Science* 47, no. 7 (September 1): 1136–1139.

de Klerk, J.N., M. Quan, and J.D. Grewar. 2020. Socio-Economic Impacts of Working Horses in Urban and Peri-Urban Areas of the Cape Flats, South Africa. *Journal of the South African Veterinary Association* 91: 1–11.

Knowles, R.D., F. Ferbrache, and A. Nikitas. 2020. Transport's Historical, Contemporary and Future Role in Shaping Urban Development: Re-Evaluating Transit Oriented Development. *Cities* 99: 1–11.

Kobayashi, A. 2014. The Dialectic of Race and the Discipline of Geography. *Annals of the Association of American Geographers* 104, no. 6: 1101–1115.

Kobayashi, A. 2019. Issues of "Race" and Early Radical Geography. In *Spatial Histories of Radical Geography*, ed. T. Barnes and E. Sheppard, 37–58. London: Wiley-Blackwell.

Krambeck, H.V. 2006. *The Global Walkability Index*. Cambridge: Massachusetts Institute of Technology.

Kunzig, R. 2019. To Build the Cities of the Future, We Must Get out of Our Cars. *National Geographic*, March 14. https://www.nationalgeographic.com/magazine/2019/04/to-build-cities-of-the-future-stop-driving-cars/.

Kwan, M.-P., and T. Schwanen. 2016. Geographies of Mobility. *Annals of the Association of American Geographers* 106, no. 2: 243–256.

Labuschagne, K., and H. Ribbens. 2014. Walk the Talk on the Mainstreaming of Non-Motorised Transport in South Africa. In 33rd Southern African Transport Conference in 2014 hosted in Tshwane, South Africa, 191–206. Tshwane, South Africa. https://repository.up.ac.za/handle/2263/45524

Latilla, M. 2013. Short History of Trams in Johannesburg. Johannesburg 1912 - *Suburb by Suburb Research*. https://johannesburg1912.com/2013/06/01/short-history-on-trams-in-johannesburg/.

Lauermann, J. 2016. Temporary Projects, Durable Outcomes: Urban Development through Failed Olympic Bids? *Urban Studies* 53, no. 9: 1885–1901.

Lauster, M. 2007. Walter Benjamin's Myth of the "Flâneur." *The Modern Language Review* 102, no. 1: 139–156.

Leduc, G.G. 1969. The Economic Balance Sheet of Colonialism. *Journal of Contemporary History* 4, no. 1: 37–50.

de Leeuw, S., and S. Hunt. 2018. Unsettling Decolonizing Geographies. *Geography Compass* 12, no. 7: 1–14.

Lefebvre, H. 1970. *The Urban Revolution*. Minneapolis: University of Minnesota Press.

Lemanski, C. 2009. Augmented Informality: South Africa's Backyard Dwellings as a by-Product of Formal Housing Policies. *Habitat International* 33, no. 4 (October): 472–484.

Lemon, A., ed. 1991. *Homes Apart: South Africa's Segregated Cities*. Bloomington: Indiana University Press.

Leslie, E., B. Saelens, L. Frank, N. Owen, A. Bauman, N. Coffee, and G. Hugo. 2005. Residents' Perceptions of Walkability Attributes in Objectively Different Neighbourhoods: A Pilot Study. *Health Place* 11, no. 3: 227–236.

Lewis, J. 2015. Preserving and Maintaining the Concept of Letchworth Garden City. *Planning Perspectives* 30, no. 1: 153–163.

Liebow, E. 1967. *Tally's Corner: A Study of Negro Streetcorner Men*. Washington, DC: Rowman & Littlefield.

Lipsitz, G. 2011. *How Racism Takes Place*. Philadelphia, PA: Temple University Press.

Litman, T. 2003. Economic Value of Walkability. *Transportation Research Record* 1828, no. 1: 3–11.

Litman, T. 2015. *Evaluating Complete Streets: The Value of Designing Roads for Diverse Modes, Users and Activities*. Vancouver: Victoria Transport Policy Institute.

Lo, R.H. 2009. Walkability: What Is It? *Journal of Urbanism* 2, no. 2: 145–166.

Lomp, S. 2017. *Things That Go*. New York, NY: Workman Kids.

London, J. 1903. *The People of the Abyss*. Boston, USA: L. C. Page and Company.

Loots, A. 2019. *Great Walk: Assessment of NMT Interventions by the JDA*. Johannesburg: Johannesburg Development Agency.

Lorimer, J. 2010. Elephants as Companion Species: The Lively Biogeographies of Asian Elephant Conservation in Sri Lanka. *Transactions of the Institute of British Geographers* 35, no. 4: 491–506.

Lovell, H. 2017. Policy Failure Mobilities. *Progress in Human Geography* 43, no. 1: 46–63.

Lucas, K. 2011. Making the Connections Between Transport Disadvantage and the Social Exclusion of Low Income Populations in the Tshwane Region of South Africa. *Journal of Transport Geography* 19: 1320–1334.

Lucas, K. 2018. Editorial for Special Issue of European Transport Research Review: Transport Poverty and Inequalities. *European Transport Research Review* 10, no. 17: 1–3.

Lucas, K., J. Blanco, A. Schafran, E. Verlinghieri, and R. Apaolazae. 2018. Contested Mobilities in the Latin American Context. *Journal of Transport Geography* 67: 73–75.

Lucas, K., B. Van Wee, and K. Maat. 2016. A Method to Evaluate Equitable Accessibility: Combining Ethical Theories and Accessibility-Based Approaches. *Transportation* 43: 473–490.

Luke, R. 2020. The Impact of COVID-19 on Transport in South Africa. *Journal of Transport and Supply Chain Management* 14, no. 1 (December 9): 5.

Lulka, D. 2008. The Paradoxical Nature of Growth in the US Bison Industry. *Journal of Cultural Geography* 25, no. 1: 31–56.

Luthuli, W. 2010. *Angel of the North*.

Madera, J. 2015. *Black Atlas: Geography and Flow in Nineteenth-Century African American Literature*. Durham, North Carolina: Duke University Press.

Maja, P. 2020. First Case of Covid-19 Coronavirus Reported in SA. https://www.gov.za/speeches/health-updates-coronavirus-10-mar-2020-0000.

Makhaye, C., and N. Mkhize. 2021. KwaZulu-Natal's Battered and Bruised Tourism Sector Prays for a Bumper Festive Season. *Daily Maverick*, December 15. https://www.dailymaverick.co.za/article/2021-12-15-kwazulu-natals-battered-and-bruised-tourism-sector-prays-for-a-bumper-festive-season/.

Malan, M. 2021. The Joke's on Us, South Africa. The Cruel Logic of Omicron Travel Bans — Debunked. *Daily Maverick*, December 7. https://www.dailymaverick.co.za/article/2021-12-07-the-jokes-on-us-south-africa-the-cruel-logic-of-omicron-travel-bans-debunked/.

Manaugh, K., and A. El-Geneidy. 2011. Validating Walkability Indices: How Do Different Households Respond to the Walkability of Their Neighborhood? *Transportation Research Part D* 16: 309–315.

Mangcu, X. 2016. Decolonizing South African Sociology: Building on a Shared "Text of Blackness." *Du Bois Review: Social Science Research on Race* 13, no. 1: 45–59.

Maputla, T. 2019. Photographer Kgomotso Neto Turns Taxi Rank into Art Space. *SowetanLIVE*, October 31. https://www.sowetanlive.co.za/s-mag/2019-10-31-photographer-kgomotso-neto-turns-taxi-rank-into-art-space/.

Markewicz, A., L.J. Linders, and G.Y. Moodley. 2011. Translating Ethekwini's Vision of Compact City into Reality through Integrated Transportation and Land Use Planning. In *Proceedings of the 30th Southern African Transport Conference*, Tshwane, South Africa. 1–11.

Marquardt, A., B. Blakemore, and R. Eichenholz. 2010. Stray Dogs Master Complex Moscow Subway System. *ABC News*, March 19. https://abcnews.go.com/International/Technology/stray-dogs-master-complex-moscow-subway-system/story?id=10145833.

Marschall, S. 2008. Transforming Symbolic Identity: Wall Art and the South African City. *African Arts* 41, no. 2: 12–23.

Masaoe, E. 2017. Safety of Vulnerable Road Users on a Road in Kinondoni Municipality, Dar Es Salaam, Tanzania. In *Non-Motorized Transport Integration into Urban Transport*

Planning in Africa, ed. W.V. Mitullah, M. Vanderschuren, and M. Khayesi. London: Routledge. 80–89.

Massé, F., and E. Lunstrum. 2016. Accumulation by Securitization: Commercial Poaching, Neoliberal Conservation, and the Creation of New Wildlife Frontiers. *Geoforum* 69: 227–237.

Mateo-Babiano, I. 2016. Indigeneity of Transport in Developing Cities. *International Planning Studies* 21, no. 2: 132–147.

Matlon, J. 2014. Narratives of Modernity, Masculinity, and Citizenship Amid Crisis in Abidjan's Sorbonne. *Antipode* 46, no. 3: 717–735.

Maylam, P. 1995. Explaining the Apartheid City: 20 Years of South African Urban Historiography. *Journal of Southern African Studies* 21, no. 1 (March): 19–38.

Mbembe, A., and S. Nuttall. 2004. Writing the World from an African Metropolis. *Public Culture* 16, no. 3: 347–372.

McCann, B., and S. Rynne. 2017. *Complete Streets: Best Policy and Implementation Practices*. Washington, DC: American Planning Association.

McCann, E. 2011. Urban Policy Mobilities and Global Circuits of Knowledge: Toward a Research Agenda. *Annals of Association of American Geographers* 101, no. 1: 107–130.

McCann, E., and K. Ward. 2012. Assembling Urbanism: Following Policies and "Studying through" the Sites and Situations of Policy Making. *Environment and Planning A* 44, no. 1: 42–51.

McCann, E., and K. Ward. 2015. Thinking through Dualisms in Urban Policy Mobilities. *International Journal of Urban and Regional Research* 39, no. 4: 828–830.

McCann, E., and K. Ward, eds. 2011. *Mobile Urbanism: Cities and Policymaking in the Global Age*. Minneapolis: University of Minnesota Press.

McCaul, C. 1990. *No Easy Ride: The Rise and Future of the Black Taxi Industry*. Johannesburg: South African Institute of Race Relations.

McCaul, C. 2011. Negotiating the Deal to Enable the First Rea Vaya Bus Company: Agreements, Experiences and Lessons. In The 30th Southern African Transport Conference in Tshwane, South Africa. Tshwane. https://repository.up.ac.za/handle/2263/17371.

McCool, A. 2016. A Walk to Freedom: Can Joburg's Bridges Heal the Urban Scars of Apartheid? *The Guardian*, December 16, sec. Cities.

McDowell, L. 2016. Reflections on Feminist Economic Geography: Talking to Ourselves? *Environment and Planning A* 48, no. 10: 2093–2099.

McFarlane, C. 2021. Repopulating Density: COVID-19 and the Politics of Urban Value. *Urban Studies* (June 9).

McGregor, J. 2005. Crocodile Crimes: People versus Wildlife and the Politics of Postcolonial Conservation on Lake Kariba, Zimbabwe. *Geoforum* 36, no. 3: 353–369.

McKittrick, K. 2006. *Demonic Grounds: Black Women and the Cartographies of Struggle*. Minneapolis: University of Minnesota Press.

McKittrick, K. 2011. On Plantations, Prisons, and a Black Sense of Place. *Social & Cultural Geography* 12, no. 8: 947–963.

McKittrick, K., and C. Woods. 2007. No One Knows the Mysteries at the Bottom of the Ocean. In *Black Geographies and the Politics of Place*, ed. K. McKittrick and C. Woods, 1–13. Cambridge, MA: South End Press.

Mead, N.V. 2016. Can Johannesburg Reinvent Itself as Africa's First Cycle-Friendly Megacity? *The Guardian*, May 24.

Merriam-Webster Dictionary. 2016. Pedestrian. *Merriam-Webster Dictionary*. Merriam Webster.

van der Merwe, H., G. Visser, and R. Donaldson. 2016. Stellenbosch: The Birthplace of Geography Teaching at South African Universities. In *The Origin and Growth of Geography as a Discipline at South Africa Universities*, ed. G. Visser, R. Donaldson, and C. Seethal, 9–32. Stellenbosch: Sun Press.

Michael Stasik and Sidy Cissokho. 2018. Introduction to Special Issue: Bus Stations in Africa. *Africa Today* 65, no. 2: vii.

Mignolo, W. 2002. The Geopolitics of Knowledge and the Colonial Difference. *South Atlantic Quarterly* 101, no. 1: 57–96.

Mignolo, W. 2007a. INTRODUCTION: Coloniality of Power and De-Colonial Thinking. *Cultural Studies* 21, no. 2–3: 155–167.

Mignolo, W. 2007b. Delinking: The Rhetoric of Modernity, the Logic of Coloniality and the Grammar of de-Coloniality. *Cultural Studies* 21, no. 3: 449–514.

Minty, Z. 2006. Post-Apartheid Public Art in Cape Town: Symbolic Reparations and Public Space. *Urban Studies* 43, no. 2: 421–440.

Minty, Z. 2023. Urban Cultural Governance and the South African City: The Case of Johannesburg's Newtown Cultural Precinct. PhD dissertation, Cape Town: University of Cape Town.

Mitchell, W.J.T. 2006. *What Do Pictures Really Want?* Chicago, IL: University of Chicago Press.

Mitullah, W.V., and R. Opiyo. 2017. Non-Motorized Transport Infrastructure Provision on Selected Roads in Nairobi. In *Non-Motorized Transport Integration into Urban Transport Planning in Africa*, ed. W.V. Mitullah, M. Vanderschuren, and M. Khayesi. London: Routledge. 90–111.

Mitullah, W.V., M. Vanderschuren, and M. Khayesi, eds. 2017. *Non-Motorized Transport Integration into Urban Transport Planning in Africa*. London: Routledge.

Mndebele, M. 2019. How South Africa's Minibus Taxi Industry Exploits Its Drivers. *The Mail & Guardian*, December 20. https://mg.co.za/article/2019-12-20-00-how-south-africas-minibus-taxi-industry-exploits-its-drivers/.

Mobius, C. 1894. The Geographical Distribution and Habits of Whales. *The Geographical Journal* 4, no. 3: 266.

Mockaitis, T. 2020. What We Can Learn from Past Pandemics. *The Hill*. https://thehill.com/opinion/white-house/487111-what-we-can-learn-from-past-pandemics.

Mogaji, E. 2020. Impact of COVID-19 on Transportation in Lagos, Nigeria. *Transportation Research Interdisciplinary Perspectives* 6 (July 1): 100154.

Montero, S. 2016. Worlding Bogota's Ciclovia: From Urban Experiment to International "Best Practice." *Latin American Perspectives* 44, no. 2: 111–131.

Montero, S. 2018. Leveraging Bogotá: Sustainable Development, Global Philanthropy and the Rise of Urban Solutionism. *Urban Studies*: 004209801879855.

Montgomery, B., and P. Roberts. 2008. *Walk Urban: Demand, Constraints and Measurement of the Urban Pedestrian Environment*. Washington, DC: World Bank.

Moore, L. 2011. The Neoliberal Elephant: Exploring the Impacts of the Trade Ban in Ivory on the Commodification and Neoliberalisation of Elephants. *Geoforum* 42, no. 1: 51–60.

Moore, S. 2013. What's Wrong with Best Practice? Questioning the Typification of New Urbanism. *Urban Studies* 50, no. 11: 2371–2387.

Morgan, N. 2017. Cycling Infrastructure and the Development of a Bicycle Commuting Socio-Technical System: The Case of Johannesburg. *Applied Mobilities* 4, no. 1: 106–123.

Morgan, N., and B. Rink. 2023. The Past in Present African Urban Mobility Systems: Towards a Mobilities Longue Durée. *Urban Forum* 34, no. 4 (December): 413–418.

Motsoere, G. 2023. Rea Vaya Bus Operator Placed under Business Rescue. *Eyewitness News*, December 25. https://ewn.co.za/2023/12/26/rea-vaya-bus-operator-placed-under-business-rescue.

Murray, L. 2014. Reading the Mobile City Through Street Art: Belfast's Murals. In *Researching and Representing Mobilities*, ed. L. Murray and S. Upstone, 99–128. London: Palgrave Macmillan UK.

Mutongi, K. 2017. *Matatu: A History of Popular Transportation in Nairobi*. Chicago, IL: University of Chicago Press.

Naidoo, S. 2021. Local Artists Turns Taxis into Her Own Canvases in Public Art Initiative. *IOL*, May 15. https://www.iol.co.za/weekend-argus/news/local-artists-turns-taxis-into-her-own-canvases-in-public-art-initiative-bb406736-0692-4530-834b-1452097cfd64.

National Department of Transport. 1996. *White Paper on National Transport Policy*. Pretoria: Department of Transport.

National Department of Transport. 2013. *National Household Travel Survey*. Tshwane, South Africa: Department of Transport.

National Planning Commission. 2011. *Diagnosis Overview*. Pretoria: National Planning Commission in the Presidency.

National Treasury. 2019. *2019 Budget Review*. Tshwane, South Africa: National Treasury.

Naybor, D., J.P.H. Poon, and I. Casas. 2016. Mobility Disadvantage and Livelihood Opportunities of Marginalized Widowed Women in Rural Uganda. *Annals of the American Association of Geographers* 106, no. 2: 404–412.

Ndlovu-Gatsheni, S.J. 2018. *Epistemic Freedom in Africa: Deprovincialization and Decolonization*. Rethinking Development. London; New York: Routledge, Taylor & Francis Group.

Nelson, R.J. 2023. The Spatial and Social Logic of the Minibus Taxi Network: How Access May Support Social Inclusion in Cape Town, South Africa. *Applied Mobilities* 8, no. 1 (January 2): 1–25.

Newbigin, M. 1913. *Animal Geography: The Faunas of the Natural Regions of the Globe*. Oxford: Clarendon Press.

Newman, M. 1991. FOCUS; A Transit-Oriented Approach to Suburbia. *The New York Times*, November 10, sec. Real Estate. https://www.nytimes.com/1991/11/10/realestate/focus-a-transit-oriented-approach-to-suburbia.html.

Nicolau, M.D., and N.C. Davis. 2002. Restructuring South African Geography. *South African Geographical Journal* 84, no. 1: 12–20.

Njoh, A.J. 1997. Colonial Spatial Development Policies, Economic Instability, and Urban Public Transportation in Cameroon. *Cities* 14, no. 3: 133–143.

Njoh, A.J. 2007. *Planning Power: Town Planning and Social Control in Colonial Africa*. London: UCL Press.

Nkosi, O. 2020. How Covid-19 Is Impacting on Public Transport. *New Frame*, July 3. https://www.newframe.com/how-covid-19-is-impacting-on-public-transport/.

Null, S., and H. Smith. 2020. *COVID-19 Could Affect Cities for Years: Here Are 4 Ways They're Coping Now*. Washington, DC: World Resources Institute. https://www.wri.org/insights/covid-19-could-affect-cities-years-here-are-4-ways-theyre-coping-now.

Nyamai, D.N. 2021. Urban Mobility and Spatial Justice: Prospects of Non-Motorized Transport in Nairobi. In *Transport Planning and Mobility in Urban East Africa*, ed. N. Appelhans, W. Scholz, and S. Baumgart, 31–42. London: Routledge.

Nyman, J. 2003. *Postcolonial Animal Tale from Kipling to Coetzee*. Delhi, India: Atlantic Publishers & Dist.

Nyovest, C. 2013. *Cassper Nyovest Ft. OkMalumKoolKat "Gusheshe."* Video. https://www.youtube.com/watch?v=Yihc_E0tv1g.

Okyere, S.A., L.K. Frimpong, S.L. Mensah, D. Oviedo, R.O. Amoako-Sakyi, M.J. Nieto-Combariza, and M. Kita. 2023. Planning for Walkable Cities in Africa: Co-Producing Knowledge on Conditions, Practices, and Strategies. *Societal Impacts* 1, no. 1–2 (December): 100005.

Olvera, D., D. Plat, P. Pochet, and M. Sahabana. 2010. Motorized Two-Wheelers in Sub-Saharan African Cities: Public and Private Use. In *12th World Conference on Transport Research*, 1–17. Lisbon, Portgular. https://www.wctr2010.info/WCTR_General/documents/02223.pdf.

Olvera, L.D., A. Guézéré, D. Plat, and P. Pochet. 2016. Earning a Living, but at What Price? Being a Motorcycle Taxi Driver in a Sub-Saharan African City. *Journal of Transport Geography* 55: 165–174.

Onyango, G.M., and F.O. Owino. 2021. Transit Oriented Development in Medium Cities in Africa: Experiences from Kisumu, Kenya. *Journal of Geography and Regional Planning* 14, no. 2: 91–104.

Orderson, C. 2024. Thirty Years Waiting for a House: South Africa's "Backyard" Dwellers. *Al Jazeera*, April 10. https://www.aljazeera.com/features/2024/4/10/thirty-years-waiting-for-a-house-south-africas-backyard-dwellers.

Orwell, G. 1937. *The Road to Wigan Pier*. London: Left Book Club.

Oswin, N. 2020. An Other Geography. *Dialogues in Human Geography* 10, no. 1: 9–18.

Oviedo, D., S.A. Okyere, M. Nieto, M. Kita, L.F. Kusi, Y. Yusuf, and B. Koroma. 2021. Walking off the Beaten Path: Everyday Walking Environment and Practices in Informal Settlements in Freetown. *Research in Transportation Business & Management* 40 (September): 100630.

Owens, M., and J. Wolch. 2017. Lively Cities: People, Animals, and Urban Ecosystems. In *Oxford Handbook of Animal Studies*, ed. L. Kalof, 542–570. Oxford: Oxford University Press.

Oxford English Dictionary. 2012. Pedestrian. *Oxford English Dictionary*. Oxford: Oxford University Press.

Oyeyemi, A., J. Sallis, B. Deforche, A. Oyeyemi, I. de Bourdeaudhuij, and D. Van Dych. 2013. Evaluation of the Neighborhood Environment Walkability Scale in Nigeria. *International Journal of Health Geographies* 12–16.

Parker, A., and M. Rubin. 2023. Mobility Intersections: Gender, Family, Culture and Location in the Gauteng City-Region. *Urban Forum* 34, no. 4 (December): 463–479.

Parks, V. 2016. Rosa Parks Redux: Racial Mobility Projects on the Journey to Work. *Annals of the American Association of Geographers* 106, no. 2: 292–299.

Parnell, S. 2020. The Enabling Conditions of Post-Pandemic City Government. *Environment and Planning B: Urban Analytics and City Science* 47, no. 7 (September 1): 1143–1145.

Payne, S. 2020. Covid-19 Shutdown of Trains Sees Rail Agency Lose Almost R200m in Revenue. *Daily Maverick*, May 7. https://www.dailymaverick.co.za/article/2020-05-07-covid-19-shutdown-of-trains-sees-rail-agency-lose-almost-r200m-in-revenue/.

Peck, J., and N. Theodore. 2010. Mobilizing Policy: Models, Methods, and Mutations. *Geoforum* 41, no. 2. Mobilizing Policy (March): 169–174.

Peck, J., and N. Theodore. 2015. *Fast Policy: Experimental Statecraft at the Thresholds of Neoliberalism*. Minneapolis, Minnsesota: University of Minnesota Press.

Pendakur, V.S. 2005. *Non-Motorized Transport in African Cities*. SSATP Working Paper No. 80. Washington, DC: World Bank.

Pereira, R.H.M., T. Schwanen, and D. Banister. 2017. Distributive Justice and Equity in Transportation. *Transport Reviews* 37, no. 2: 170–191.

Perryer, S., D. Koloane, and T. Mosaka, eds. 2004. *10 Years, 100 Artists: Art in a Democratic South Africa*. Cape Town: Bell-Roberts.

Petersen, T. 2017. Cape Town's Dancing Taxi Guardjie on the Move. *News24*, October 6, sec. News24. https://www.news24.com/news24/cape-towns-dancing-taxi-guardjie-on-the-move-20171005.

Phadi, M. 2021. The Multiple Consciousness of Blackness: Race and Class in South Africa. *Transformation: Critical Perspectives on Southern Africa* 105, no. 1: 52–73.

Pharoah, T., and K. Buchan. 1997. *Putting London Back on Its Feet*. London: London Planning Advisory Committee.

Phillips, H. 1987. The Local State and Public Health Reform in South Africa: Bloemfontein and the Consequences of the Spanish 'flu Epidemic of 1918. *Journal of Southern African Studies* 13, no. 2: 210–233.

Phillips, H. 1988. South Africa's Worst Demographic Disaster: The Spanish Influenza Epidemic of 1918. *South African Historical Journal* 20, no. 1: 57–73.

Phillips, H. 2001. AIDS in the Context of South Africa's Epidemic History: Preliminary Historical Thoughts. *South African Historical Journal* 45, no. 1: 11–26.

Phillips, H. 2012. *Epidemics: The Story of South Africa's Five Most Lethal Human Diseases*. Athens, Ohio: Ohio University Press.

Phillips, R.E. 1935. Readers' Points of View. Rand Daily Mail, January.

Philo, C., and C. Wilbert. 2000. *Animal Spaces, Beastly Places: New Geographies of Human-Animal Relations*. London: Routledge.

Philo, C., and J. Wolch. 1998. Through the Geographical Looking Glass: Space, Place and Society-Animal Relations. *Society & Animals* 6, no. 2: 103–118.

Philol, C. 1995. Animals, Geography, and the City: Notes on Inclusions and Exclusions. *Environment and Planning D: Society and Space* 13, no. 6: 655–681.

Pierre, J. 2013. *The Predicament of Blackness: Postcolonial Ghana and the Politics of Race*. Chicago, IL: The University of Chicago Press.

Pikoli, B.C.S. and Z. 2020. Rail, Bus and Taxi Organisations Scramble as They Mobilise to Fight Covid-19. *Daily Maverick*. https://www.dailymaverick.co.za/article/2020-03-18-rail-bus-and-taxi-organisations-scramble-as-they-mobilise-to-fight-covid-19/.

Pirie, G. 1983. Urban Bus Boycott in Alexandra Township, 1957. *African Studies* 42, no. 1: 67–77.

Pirie, G. 1986. Johannesburg Transport 1905-1945: African Capitulation and Resistance. *Journal of Historical Geography* 12: 42–55.

Pirie, G. 1989. Racial Segregation on Johannesburg Trams: Procedures and Protest, 1904–1927. *African Studies* 48, no. 1: 37–54.

Pirie, G. 1992. Rolling Segregation into Apartheid: South African Railways, 1948-53. *Journal of Contemporary History* 27, no. 4: 671–693.

Pirie, G. 2013. Transport Geography in South Africa. *Journal of Transport Geography* 31: 312–314.

Pirie, G. 2014. Transport Pressures in Urban Africa: Practices, Policies, Perspectives. In *Africa's Urban Revolution*, ed. S. Parnell and E. Pieterse, 133–147. London: Zed Books Ltd.

Plessy v. Ferguson heard in the US Supreme Court in 1896. Case number, the volumber number of the United States Reports - 163 U.S. 537 (1896).

https://supreme.justia.com/cases/federal/us/163/537/#:~:text=Ferguson%2C%20163%20U.S.%20537%20(1896)

Pollio, A. 2019. Forefronts of the Sharing Economy: Uber in Cape Town. *International Journal of Urban and Regional Research*. https://onlinelibrary.wiley.com/doi/abs/10.1111/1468-2427.12788.

Pollio, A., L.R. Cirolia, and J. Ong'iro Odeo. 2023. Algorithmic Suturing: Platforms, Motorcycles and the "Last Mile" in Urban Africa. *International Journal of Urban and Regional Research* 47, no. 6 (November): 957–974.

Porter, G. 2016. Mobilities in Rural Africa: New Connections, New Challenges. *Annals of the American Association of Geographers* 106, no. 2: 434–441.

Prince, R. 2016. The Spaces in between: Mobile Policy and the Topographies and Topologies of the Technocracy. *Environment and Planning D* 34, no. 3: 420–437.

Priya Uteng, T., and K. Lucas, eds. 2018. *Urban Mobilities in the Global South*. Abingdon & New York: Routledge.

Prytherch, D. 2018. *Law, Engineering and the American Right-of-Way*. London: Palgrave Macmillan.

Pulido, L. 2002. Reflections on a White Discipline. *The Professional Geographer* 54, no. 1: 42–49.

Quijano, A. 2007. Coloniality and Modernity/Rationality. *Cultural Studies* 21, no. 2–3: 168–178.

Radcliffe, S.A. 2017a. Geography and Indigeneity I: Indigeneity, Coloniality and Knowledge. *Progress in Human Geography* 41, no. 2: 220–229.

Radcliffe, S.A. 2017b. Decolonising Geographical Knowledges. *Transactions of the Institute of British Geographers* 42, no. 3: 329–333.

Radcliffe, S.A. 2018. Geography and Indigeneity II: Critical Geographies of Indigenous Bodily Politics. *Progress in Human Geography* 42, no. 3 (June 1): 436–445.

Rand Daily Mail. 1918. Spanish Influence in South Africa, September 17.

Rapoport, E., and A. Hult. 2017. The Travelling Business of Sustainable Urbanism: International Consultants as Norm-Setters. *Environment and Planning A* 49, no. 8: 1779–1796.

Ratshilumela, T. 2018. Non-Motorized Transport Infrastructure - "Complete Street" presented at the Masterclass on Transport Planning, July 25, Johannesburg.

Renne, J.L., S. Hamidi, and R. Ewing. 2016. Transit Commuting, the Network Accessibility Effect, and the Built Environment in Station Areas Across the United States. *Research in Transportation Economics* 60: 35–43.

Richards, P. 1974. Kant's Geography and Mental Maps. *Transactions of the Institute of British Geographers* no. 61: 1–16. https://www.jstor.org/stable/621596.

Richardson, H., and R. Moss. 2024. North East Mayor: What's the Role and Who Is in the Running? *BBC News*, April 18, sec. England. https://www.bbc.com/news/articles/cgl6vgyg8jdo.

Rigolon, A., Z. Toker, and N. Gasparian. 2018. Who Has More Walkable Routes to Parks? An Environmental Justice Study of Safe Routes to Parks in Neighborhoods of Los Angeles. *Journal of Urban Affairs* 40, no. 4: 576–591.

Riis, J. 1890. *How the Other Half Lives*. New York, NY: Simon & Schuster.

Rink, B. 2016. Race and the Micropolitics of Mobility: Mobile Autoethnography on a South African Bus Service. *Transfers* 6, no. 1: 62–79.

Rink, B. 2020. Capturing Amaphela: Negotiating Township Politics through Shared Mobility. *Geoforum* 136: 232-241.

Rink, B., and J. Crow. 2021. Horse/Power: Human–Animal Mobile Assemblage in the Contemporary City. *Contemporary Social Science* 16, no. 1: 84–95.

Rizzo, M. 2017. *Taken for a Ride: Grounding Neoliberalism, Precarious Labour, and Public Transport in an African Metropolis*. Oxford: Oxford University Press.

Road Traffic Management Corporation. 2018. *State of the Road Report*. Tshwane, South Africa: National Department of Transport.

Robinson, C. 1983. *Black Marxism: The Making of the Black Radical Tradition*. Chapel Hill, North Carolina: University of North Carolina Press.

Robinson, J. 2011. The Spaces of Circulating Knowledge: City Strategies and Global Urban Governmentality. In *Mobile Urbanism: Cities and Policymaking in the Global Age*, ed. E. McCann and K. Ward, 15–40. Minneapolis: University of Minnesota Press.

Robinson, J. 2015. "Arriving at" Urban Policies: The Topological Space of Urban Policy Mobility. *International Journal of Urban and Regional Research* 39, no. 4: 831–834.

Robinson, J., P. Harrison, J. Shen, and F. Wu. 2020. Financing Urban Development, Three Business Models: Johannesburg, Shanghai and London. *Progress in Planning* 154: 1–42.

Rodney, W. 1972. How Europe Underdeveloped Africa. London: Verso Books.

Rose, G. 2001. *Visual Methodologies: An Introduction to the Interpretation of Visual Materials*. London: Sage.

Rose, G. 2003. On the Need to Ask How, Exactly, Is Geography "Visual"? *Antipode* 35, no. 2: 212–221.

Rose, H. 1971. *The Black Ghetto; A Spatial Behavioral Perspective*. New York: McGraw-Hill.

Rose, H. 1976. *Black Suburbanization: Access to Improved Quality of Life or Maintenance of the Status Quo?* Pensacola, FL: Ballinger Publishing Company.

Rose, H., and P. McClain. 1990. *Race, Place and Risk: Black Homicide in Urban America*. Albany, New York: State University of New York Press.

Rose, R. 2024. South Africa Sets Ambitious Timeline to End Rolling Blackouts. *The Financial Times*, March 12. https://www.ft.com/content/362c08c3-518f-48c1-922d-e15dfeb36899.

van Ryneveld, P. 2006. *Towards a Discussion on Refining the System of Decentralization and the Division of Powers and Functions Between the Three Spheres of Government*. South Africa: Department of Provincial and Local Government.

SA Taxi. 2017. Taxis as Mobile Galleries Bring International Art Trends to South African. SA Taxi. *Taxis as Mobile Galleries*. https://sataxi.co.za/taxis-mobile-galleries/.

Said, E.W. 1978. *Orientalism*. New York: Random House.

Samaha, P., and A. Mohtar. 2020. Decoding an Urban Myth: An Inquiry into the Van Line 4 System in Beirut, Lebanon. *Journal of Transport Geography* 85: 1–9.

Santos, M. 1977. Planning Underdevelopment. *Antipode* 9, no. 3: 86–98.

Santos, M. 1979. *The Shared Space: The Two Circuits of the Urban Economy in Underdeveloped Countries*. London: Routledge.

Schalekamp, H., and R. Behrens. 2009. Engaging Paratransit on Public Transport Reform Initiatives in South Africa: A Critique of Policy and an Investigation of Appropriate Engagement Approaches. *Research in Transportation Economics* 29, no. 1, 371–378.

Schalekamp, H., R. Behrens, and P. Wilkinson. 2010. Regulating Minibus Taxis: A Critical Review of Progress and a Possible Way Forward. In The 29th Southern African Transort Conference. Tshwane, South Africa.

Schuetz, J. 2015. Do Rail Transit Stations Encourage Neighbourhood Retail Activity? *Urban Studies* 52, no. 14: 2699–2723.

Schwanen, T. 2018a. Towards Decolonised Knowledge about Transport. *Palgrave Communications* 4, no. 1: 1–6.

Schwanen, T. 2018b. Geographies of Transport III: New Spatialities of Knowledge Production? *Progress in Human Geography* 42, no. 3: 463–472.

Schwanen, T. 2020. Towards Decolonial Human Subjects in Research on Transport. Journal of Transport Geography. *Journal of Transport Geography* 88, 1–11.

Seiler, C. 2007. The Significance of Race to Transport History. *The Journal of Transport History* 28, no. 2: 307–311.

Sen, A. 2009. *The Idea of Justice*. Cambridge, MA: Harvard University Press.

Sey, J. 2012. *The People Shall Move!: A History of Public Transport*. Johannesburg, South Africa: Department of Transport.

Shabazz, R. 2015. *Spatializing Blackness: Architectures of Confinement and Black Masculinity in Chicago*. Chicago: University of Illinois Press.

Shaw, W.S., R.D.K. Herman, and G.R. Dobbs. 2006. Encountering Indigeneity: Re-Imagining and Decolonizing Geography. *Geografiska Annaler: Series B, Human Geography* 88, no. 3: 267–276.

Shearmur, R., P. Ananian, U. Lachapelle, M. Parra-Lokhorst, F. Paulhiac, D.-G. Tremblay, and A. Wycliffe-Jones. 2021. Towards a Post-COVID Geography of Economic Activity: Using Probability Spaces to Decipher Montreal's Changing Workscapes. *Urban Studies* (July 16)59, no. 10: 2053–2075. https://doi.org/10.1177/00420980211022895.

Sheller, M. 2018a. *Mobility Justice: The Politics of Movement in an Age of Extremes*. New York, NY: Verso Books.

Sheller, M. 2018b. Theorising Mobility Justice. *Tempo Social* 30, no. 2: 17–34.

Sheller, M., and J. Urry. 2006. The New Mobilities Paradigm. *Environment and Planning* 38: 207–226.

Sheller, M., and J. Urry. 2016. Mobilizing the New Mobilities Paradigm. *Applied Mobilities* 1, no. 1: 10–25.

Shields, G. 1979. *Port Elizabeth Tramways. A Short History Of Port Elizabeth's Road Passenger Transport Services 1879 - 1979*. Port Elizabeth: Privately Printed.

Shingne, M.C. 2022. The More-than-Human Right to the City: A Multispecies Reevaluation. *Journal of Urban Affairs* 44, no. 2: 137–155.

Shorten, J. 1970. *The Johannesburg Saga*. Johannesburg: John R. Shorten.

Simelane, B.C. 2020. Taxis at 70 Percent Capacity Are Now Allowed – If All Passengers Wear Surgical Masks. *Daily Maverick*, April 2. https://www.dailymaverick.co.za/article/2020-04-02-taxis-at-70-percent-capacity-are-now-allowed-if-all-passengers-wear-surgical-masks/.

Simelane, B.C. 2024. "Opportunism" — Taxi Drivers Query Timing of Covid Relief Fund Extension. *Daily Maverick*, March 18. https://www.dailymaverick.co.za/article/2024-03-18-government-opportunism-taxi-drivers-question-timing-of-covid-19-relief-fund-extension/.

Skarjune, D. 1989. Boston/Cambridge. Riding on the Red Line. Arts on the Line. *Public Art Review* 1, no. 2: 12.

Smith, L.T. 1999. *Decolonizing Methodologies: Research and Indigenous Peoples*. London: Zed Books Ltd.

Smith, N. 1984. *Uneven Development: Nature, Capital, and the Production of Space*. London: Blackwell.

Soja, E.W. 1989. *Postmodern Geographies: The Reassertion of Space in Critical Social Theory*. London: Verso.

South Africa Competition Commission. 2021. *Market Inquiry into the Land Based Public Passenger Transport Sector: Main Report*. Pretoria: Competition Commission.

South African Cities Network. 2011. *State of the Cities Report*. Johannesburg, South Africa: South African Cities Network.

South African government. 2022. Travel - Coronavirus COVID-19. https://www.gov.za/covid-19/individuals-and-households/travel-coronavirus-covid-19.

South African National Government. 1914. *Prevention of Cruelty Act*. Vol. No 8.

South African National Government. 1935. *Performing Animals Protection Act*.

South African National Government. 1962. Animal Protection Act.

Southworth, M. 2005. Designing the Walkable City. *Journal of Urban Planning and Development* 131, no. 4: 246–257.

Sovacool, B.K., C. Daniels, and A. AbdulRafiu. 2022. Transitioning to Electrified, Automated and Shared Mobility in an African Context: A Comparative Review of Johannesburg, Kigali, Lagos and Nairobi. *Journal of Transport Geography* 98 (January): 103256.

Spear, A. 1967. *Black Chicago: The Making of Negro Ghetto*. Chicago, IL: University of Chicago Press.

Spit, T., and B. Patton. 1976. *Johannesburg Tramways: A History of the Tramways of the City of Johannesburg*. London: The Light Railway Transport League.

Staff Reporter. 1934a. Council v. Tramways Manager. *Rand Daily* Mail, March.

Staff Reporter. 1934b. Natives Need Seats on Tramcars. *Rand Daily Mail*, April.

Staff Reporter. 1957a. The "Sympathy" Boycott. *Rand Daily Mail*, February.

Staff Reporter. 1957b. It Started on January Seven. *Rand Daily Mail*, March.

Staff Reporter. 1957c. Boycott: All Natives May Be Riding Today. *Rand Daily Mail*, April.

Staff Reporter. 1979. Mounting Costs of Johannesburg's Segregated Buses. *Rand Daily Mail*, October.

Stats SA. 2021a. Steep Slump in GDP as COVID-19 Takes Its Toll on the Economy. https://www.statssa.gov.za/?p=13601.

Stats SA. 2021b. Quarterly Labour Force Survey (QLFS) – Q3:2021. https://www.statssa.gov.za/?p=14957.

Stent, J., M.-A. Gontsana, and T. Washinyira. 2021. How Risky Is Public Transport for Infectious Disease? We Took a Series of Trips to Put It to the Test. *Daily Maverick*, June 29. https://www.dailymaverick.co.za/article/2021-06-29-how-risky-is-public-transport-for-infectious-disease-we-took-a-series-of-trips-to-put-it-to-the-test/.

de Suremain, M.-A. 2019. Between Political Reserve and Scientific Change: French Geographers and Decolonisation in Tropical Africa, 1945-1967. *Journal of Historical Geography* 66: 9–19.

Sustrans. 2022. *Walking and Cycling Index 2021: Tyneside*.

Sustrans. 2024. Transport Hustings for North East Mayor Candidates, March 28. https://www.sustrans.org.uk/our-blog/news/2024/march/transport-hustings-for-north-east-mayor-candidates.

Suzuki, H., R. Cervero, and K. Iuchi. 2013. *Transforming Cities with Transit: Transit and Land-Use Integration for Sustainable Urban Development*. Washington, DC: World Bank.

Swanson, M.W. 1977. The Sanitation Syndrome: Bubonic Plague and Urban Native Policy in the Cape Colony, 1900-1909. *Journal of African History* 18, no. 3: 387–410.

Taaffe, E.J., R.L. Morrill, and P.R. Gould. 1963. Transport Expansion in Underdeveloped Countries: A Comparative Analysis. *Geographical Review* 53, no. 4: 503–529.

Tau, P. 2013. State of the City Address May 9, Johannesburg, South Africa.

Temenos, C., and J. Lauermann. 2020. The Urban Politics of Policy Failure. *Urban Geography* 41, no. 9: 1109–1118.

The Economist. 2017. South Africa Has One of the World's Worst Education Systems. *The Economist*, January 7. https://www.economist.com/middle-east-and-africa/2017/01/07/south-africa-has-one-of-the-worlds-worst-education-systems.

The Star. 1977. Let's Get Moving – Over to the Network. *The Star*, July 29.

Theodore, N., and J. Peck. 2011. Framing Neoliberal Urbanism: Translating "Common Sense" Urban Policy across the OECD Zone. *European Urban and Regional Studies* 19, no. 1: 20–41.

wa Thiong'o, N. 1981. *Decolonising the Mind: The Politics of Language in African Literature*. Nairobi: East Africa Educational Publishers.

Thomas, D.P. 2013. The Gautrain Project in South Africa: A Cautionary Tale. *Journal of Contemporary African Studies* 31, no. 1: 77–94.

Thomson, M. 2007. Placing the Wild in the City: "Thinking with" Melbourne's Bats. *Society & Animals* 15, no. 1: 79–95.

Todes, A., and J. Robinson. 2020. Re-Directing Developers: New Models of Rental Housing Development to Re-Shape the Post-Apartheid City? *Environment and Planning A: Economy and Space* 52, no. 2: 297–317.

Tognotti, E. 2013. Lessons from the History of Quarantine, from Plague to Influenza A. *Emerging Infectious Diseases* 19, no. 2: 254–259.

Toivanen, A.-L. 2021. *Mobilities and Cosmopolitanisms in African and Afrodiasporic Literatures*. Leiden, The Netherlands: Brill.

Transport Department. 2016. Integrated Transport Network and Implications for the City presented at the Economic Growth, Technical EMT, March 23, Johannesburg.

Trubina, E. 2020. Sidewalk Fix, Elite Maneuvering and Improvement Sensibilities: The Urban Improvement Campaign in Moscow. *Journal of Transport Geography* 83: 1–9.

Tuck, E., and K.W. Yang. 2012. Decolonization Is Not a Metaphor. *Decolonization: Indigeneity, Education & Society* 1, no. 1: 1–40.

Tumlin, J., A. Millard-Ball, C. Zucker, and P. Siegman. 2003. How to Make Transit-Oriented Development Work. *Planning* 69: 14–19.

Turner, M.D. 2009. Capital on the Move: The Changing Relation between Livestock and Labor in Mali, West Africa. *Geoforum* 40, no. 5: 746–755.

Tyilo, M. 2021. Taxis and Art: Rethinking "Cape Malay" and "Coloured" Identity. *Daily Maverick*, June 1. https://www.dailymaverick.co.za/article/2021-06-01-taxis-and-art-rethinking-cape-malay-and-coloured-identity/.

Uddin, L. 2015. *Zoo Renewal: White Flight and the Animal Ghetto*. Minneapolis, Minnesota: University of Minnesota Press.

United Nations. 2020. *Policy Brief: COVID-19 in an Urban World.*

United Nations Environment Program and United Nations Human Settlements Program. 2022. *Walking and Cycling in Africa: Evidence and Good Practice to Inspire Action*. Nairobi.

Urbanik, J. 2012. *Placing Animals: An Introduction to the Geography of Human-Animal Relations*. Lanham, Maryland: Rowman & Littlefield.

Urbanik, J., and M. Morgan. 2013. A Tale of Tails: The Place of Dog Parks in the Urban Imaginary. *Geoforum* 44: 292–302.

US Department of Transportation. 1997. *Walkability Checklist: How Walkable Is Your Community?* Washington, DC: US Department of Transportation.

Uteng, T.P., and K. Lucas, eds. 2017. *Urban Mobilities in the Global South*. London: Routledge.

Van der Westhuizen, J. 2008. Glitz, Glamour and the Gautrain: Mega-Projects as Political Symbols. *Politikon* 34, no. 3: 333–351.

Van Onselen, C. 1979. The World the Mineowners Made: Social Themes in the Economic Transformation of the Witwatersrand, 1886-1914. *Review* 3, no. 2: 289–323.

Van Onselen, C. 2001. *New Babylon New Nineveh: Everyday Life on the Witwatersrand, 1886-1914*. Johannesburg: Jonathan Ball Publishers.

Van Sittert, L., and S. Swart. 2003. Canis Familiaris: A Dog History of South Africa. *South African Historical Journal* 48, no. 1: 138–173.

Van Wee, B., and K. Geurs. 2011. Discussing Equity and Social Exclusion in Accessibility Evaluations. *European Journal of Transport and Infrastructure Research* 11, no. 4: 350–367.

Venkile, L. 2021. Transport on COVID-19 Taxi Relief Support Fund. South African Government Media Statement. https://www.gov.za/speeches/transport-covid-19-taxi-relief-support-fund-20-dec-2021-0000.

Venter, C. 2013. The Lurch Towards Formalisation: Lessons from the Implementation of BRT in Johannesburg, South Africa. *Research in Transportation Economics* 39: 114–120.

Venter, I. 2018. No Longer Flavour of the Month, SA Rethinks Its Bus Rapid Transit Systems. *Engineering News*, July 27. https://www.engineeringnews.co.za/print-version/no-longer-flavour-of-the-month-sa-rethinks-its-bus-rapid-transit-systems-2018-07-27-1.

Verlinghieri, E., and J. Middleton. 2020. Decolonising and Provincializing Knowledge within the Neoliberal University? The Challenge of Teaching about Sustainable Transport. *Journal of Transport Geography* 88: 1–9.

Verlinghieri, E., and T. Schwanen. 2020. Transport and Mobility Justice: Evolving Discussions. *Journal of Transport Geography* 87: 102798.

Visser, G., R. Donaldson, and C. Seethal, eds. 2016. *The Origin and Growth of Geography as a Discipline at South Africa Universities*. Stellenbosch: Sun Press.

Vladislavić, I. 2006. *Portrait with Keys: The City of Johannesburg Unlocked*. London: Portobello Books.

Walters, J. 2013. Overview of Public Transport Policy Developments in South Africa. *Research in Transportation Economics* 39, no. 1: 34–45.

Ward, K. 2006. "Policies in Motion", Urban Management and State Restructuring: The Trans-Local Expansion of Business Improvement Districts. *International Journal of Urban and Regional Research* 30, no. 1 (March): 54–75.

Watson, V. 2009. Seeing from the South: Refocusing Urban Planning on the Globe's Central Urban Issues. *Urban Studies* 46, no. 11: 2259–2275.

Watt, F. 2020. *"That's Not My Bus" by Fiona Watt*. London: Usbourne Publishing. https://usborne.com/gb/that-s-not-my-bus-9781474972130.

Webster, D. 2020a. Part One: The 1918 Flu Pandemic to Remember. New Frame, September 17. https://www.newframe.com/part-one-the-1918-flu-pandemic-to-remember/.

Webster, D. 2020b. Part Two: The 1918 Flu Pandemic That Made Cities. New Frame, September 18. https://www.newframe.com/part-two-the-1918-flu-pandemic-that-made-cities/.

Welford, M. 2018. *Geographies of Plague Pandemics: The Spatial-Temporal Behavior of Plague to the Modern Day*. Oxford: Routledge.

Wells, K. 2014. Policyfailing: The Case of Public Property Disposal in Washington, D.C. *ACME* 13, no. 3: 473–494.

Wesso, H. 1992. *The Colonisation of the Geographical Mind: A Critical Contextual Analysis of the Institutionalisation and Establishment of Geography as an Academic Discipline in South Africa*. Johannesburg: University of the Witwatersrand.

WHO. 2015. Factors That Contributed to Undetected Spread of the Ebola Virus and Impeded Rapid Containment. https://www.who.int/news-room/spotlight/one-year-into-the-ebola-epidemic/factors-that-contributed-to-undetected-spread-of-the-ebola-virus-and-impeded-rapid-containment.

WHO. 2020a. Updated WHO Recommendations for International Traffic in Relation to COVID-19 Outbreak. https://www.who.int/news-room/articles-detail/updated-who-recommendations-for-international-traffic-in-relation-to-covid-19-outbreak.

WHO. 2020b. WHO Timeline - COVID-19. https://www.who.int/news/item/27-04-2020-who-timeline---covid-19.

WHO. 2021. WHO Advice for International Traffic in Relation to the SARS-CoV-2 Omicron Variant (B.1.1.529). https://www.who.int/news-room/articles-detail/who-advice-for-international-traffic-in-relation-to-the-sars-cov-2-omicron-variant.

Whol, M., and B. Martin. 1967. *Traffic System Analysis for Engineers and Planners*. New York: McGraw-Hill.

Wilhelm-Solomon, M. 2020. The City Otherwise: The Deferred Emergency of Occupation in Inner-City Johannesburg. *Cultural Anthropology* 35, no. 3: 404–434.

Wilkinson, A., and M. Leach. 2015. Briefing: Ebola–Myths, Realities, and Structural Violence. *African Affairs* 114, no. 454: 136–148.

Wilkinson, P. 2006. "Transit Oriented Development": A Strategic Instrument for Spatial Restructuring and Public Transport System Enhancement in South African Cities? In *Proceedings of the 25th Southern African Transport Conference*, 1–11. Tshwane, South Africa.

Wilkinson, P., and A. Marks. 2007. Promoting Transit Oriented Development in the Atlantis Corridor, Cape Town: Towards an Implementable Model. In *Proceedings of the 26th Southern African Transport Conference*, 1–12. Tshwane, South Africa.

Williams, M. 2003. Klipfontein Bus Plan "Poverty-Busting Driving Force." *Cape Argus*, April 23.

Wilson, B. 1989. Birmingham Segregation: Is It a Product of Black-White Socioeconomic Differences? *Southeastern Geographer* 29, no. 2: 922–105.

Wilson, B. 2000. *Race and Place in Birmingham: The Civil Rights and Neighborhood Movements*. Lanham, MD: Rowman & Littlefield Publishers.

Wilson, B. 2002. Critically Understanding Race-Connected Practices: A Reading of W. E. B. Du Bois and Richard Wright. *The Professional Geographer* 54, no. 1: 31–41.

Wilson, R.T. 2018. Domestic Livestock in African Cities: Production, Problems and Prospects. *Open Urban Studies and Demography Journal* 4, no. 1: 1–14.

Wohlwill, D.E. 1996. Development along a Busway: A Case Study of Development along the Martin Luther King, Jr. East Busway in Pittsburgh, Pennsylvania. In *Proceedings of the 1996 Rapid Transit Conference of the American Public Transit Association*. Atlanta, Georgia.

Wolch, J. 1998. Zoopolis. In *Animal Geographies: Place, Politics and Identity in the Nature-Culture Borderlands*, ed. J. Wolch and J. Emel, 119–138. London: Verso.

Wolch, J. 2002. Anima Urbis. *Progress in Human Geography* 26, no. 6: 721–742.

Wolch, J., and J. Emel. 1995. Guest Editorial: Bringing Animals Back In. *Environment and Planning D: Society and Space* 13, no. 6: 632–636.

Wolch, J., and J. Emel. 1998. *Animal Geographies: Place, Politics and Identity in the Nature-Culture Borderlands*. London: Verso.

Wolch, J., K. West, and T. Gaines. 1995. Transspecies Urban Theory. *Environment and Planning D: Society and Space* 13, no. 6: 735–760.

Wolf, M. 2016. Rethinking Urban Epidemiology: Natures, Networks and Materialities. *International Journal of Urban and Regional Research* 40, no. 5: 958–982.

Wolf, P.J., and F. Hamilton. 2022. Managing Free-Roaming Cats in U.S. Cities: An Object Lesson in Public Policy and Citizen Action. *Journal of Urban Affairs* 44, no. 2: 221–242.

Wood, A. 2014. Moving Policy: Global and Local Characters Circulating Bus Rapid Transit through South African Cities. *Urban Geography* 35, no. 8: 1238–1254.

Wood, A. 2015a. Multiple Temporalities of Policy Circulation: Gradual, Repetitive and Delayed Processes of BRT Adoption in South African Cities. *International Journal of Urban and Regional Research* 39, no. 3: 568–580.

Wood, A. 2015b. The Politics of Policy Circulation: Unpacking the Relationship between South African and South American Cities in the Adoption of Bus Rapid Transit. *Antipode* 47, no. 4: 1062–1079.

Wood, A. 2017. Advancing Development Projects through Mega-Events: The 2010 Football World Cup and Bus Rapid Transit in South Africa. *Urban Geography* (November 27)40, no. 4: 428–444.

Wood, A. 2019a. Circulating Planning Ideas from the Metropole to the Colonies: Understanding South Africa's Segregated Cities through Policy Mobilities. *Singapore Journal of Tropical Geography* 40, no. 2: 257–271.

Wood, A. 2019b. Disentangling the Nexus of Global Intermediaries: The Case of Bus Rapid Transit. *Urban Development Issues* 62: 17–27.

Wood, A. 2020a. Tracing the Absence of Bike-Share in Johannesburg: A Case of Policy Mobilities and Non-Adoption. *Journal of Transport Geography* 83: 1–7.

Wood, A. 2020b. Decolonising Cities of the Global South in the Classroom and Beyond. *Town Planning Review* 91, no. 5: 535–552.

Wood, A. 2020c. Tracing Urbanism: Methods of Actually Doing Comparative Studies in Johannesburg. *Urban Geography* 41, no. 2: 293–311.

Wood, A. 2021. Problematizing Concepts of Transit-Oriented Development in South African Cities. *Urban Studies* 59, no. 12: 2451–2467.

Wood, A. 2022a. *How Cities Learn: Tracing Bus Rapid Transit in South Africa*. London: Wiley-Blackwell.

Wood, A. 2022b. Problematizing the Concept of Walkability in Johannesburg. *Journal of Urban Affairs* 46, no. 2: 237–251.

Wood, A. 2023. Bringing Transport into Black Geographies: Policies, Protests, and Planning in Johannesburg. *Annals of the American Association of Geographers* 113, no. 4: 1020–1033.

Wood, A., W. Kębłowski, and T. Tuvikene. 2020. Decolonial Approaches to Urban Transport Geographies: Introduction to the Special Issue. *Journal of Transport Geography* 88: 1–5.

Woods, C. 1998. *Development Arrested: The Blues and Plantation Power in the Mississippi Delta*. New York: Verso.

Woods, C. 2017. *Development Drowned and Reborn: The Blues and Bourbon Restorations in Post-Katrina New Orleans*. Athens: University of Georgia Press.

Woolf, S. 2007. *Taxi Hand Signs*. Johannesburg: Frankfurt Rights.

Woolf, S. 2013. *South African Taxi Hand Signs: Documenting the History and Significance of Taxi Hand Signs through Anthropology and Art, Including the Invention of a Tactile Shape-Language for Blind People*. Johannesburg: University of the Witwatersrand. Accessed on October 2, 2023. https://core.ac.uk/reader/39673660.

Worldometer. 2024a. South Africa COVID - Coronavirus Statistics. https://www.worldometers.info/coronavirus/country/south-africa/.

Worldometer. 2024b. COVID - Coronavirus Statistics. https://www.worldometers.info/coronavirus/.

Wright, H.K. 2016. Stuart Hall's Relevance for the Study of African Blackness. *International Journal of Cultural Studies* 19, no. 1: 85–99.

Wright, M. 2015. *Physics of Blackness: Beyond the Middle Passage Epistemology*. Minneapolis: University of Minnesota Press.

Yeo, J.-H., and H. Neo. 2010. Monkey Business: Human–Animal Conflicts in Urban Singapore. *Social & Cultural Geography* 11, no. 7: 681–699.

Yong, E. 2018. The Next Plague Is Coming. Is America Ready? *The Atlantic*. https://www.theatlantic.com/magazine/archive/2018/07/when-the-next-plague-hits/561734/.

Yue, R.P.H., H.F. Lee, and C.Y.H. Wu. 2017. Trade Routes and Plague Transmission in Pre-Industrial Europe. *Scientific Reports* 7 (October 11): 12973.

Zakwe, S. 2020. Modern Day Pantsula and Dan Carter Present: Modern Day Taxi Owner. *Zkhiphani*, October 20. https://www.zkhiphani.co.za/modern-day-pantsula-and-dan-carter-present-modern-day-taxi-owner/.

Zavetoski, S., and J. Agyeman, eds. 2015. *Incomplete Streets: Processes, Practices and Possibilities*. London: Routledge.

Zulu, S. 2019. City of Joburg Condemns Vandalism of Rea Vaya Station. *Eyewitness News*, April 23. Accessed on February 1, 2024. https://ewn.co.za/2019/04/23/city-of-joburg-condemns-vandalism-of-rea-vaya-station.

Index

Note: *Italic* page numbers refer to figures.

Africa/African: Accra 31, 36, 72, 133; colonialism 21; flâneur 82
African National Congress (ANC) 49, 130, 132
Agbiboa, D. 11, 36
Agyeman, J. 70, 71
Airport Station 110
Ake, C. 25–26
Alderman, D. 44, 45
Ali, S.H. 117, 120
Allara, P. 113
American Association of Retired Persons 70
American Planning Association 70
Anderson, B. 104
Angel of the North (Luthuli) 101
Anglophone geography 39, 42
animal geographies 15, 84; and automobility 89–90; challenges 84; conceptual and geographical othering 87; cultural 86; dichotomies 96; epistemological and methodological approaches 87; "first wave" 85; "fourth wave" 86–87; imagery and aesthetics of 95–96; and mobilities/movements 85, 89–97; policies governing 94–95; "second wave" 86; spatiality of 88; tactics for 85; "third wave" 86; and transportation 89–92; urban 87–89; visual methods 92
Animal Liberation Front 86
Animal Protection Act (1962) 94
anima urbis 88
anthropocentric approaches 19, 84
anti-capitalist approach 40
apartheid 2, 5, 9, 19, 20, 22, 25–27, 30–32, 35, 36, 40, 49, 58, 74, 76, 95, 96, 100, 110, 112, 130, 131
apartheid spatial planning 2, 9, 31, 32, 76; buses and 5–6
Appleyard, D. 70
artful cities 104
Art Gallery Station 108
art geographies 3, 10, 13, 15, 19, 98–114; *see also* transp-art-ation
Association of American Geographers (AAG) 39, 41, 42
automobility: animal geographies and 89–90

bacteriological city 117
Bear, C. 92
Beck, A.M. 86
Behrens, R. 12, 55, 56, 74
Bell, D. 42
Benjamin, W. 67
Bennett, C. 86
Benton, J.S. 72
Berg, C. van den 101
Bethlehem, L. 32, 107
Bickford, G. 56
Bicycle Empowerment Network 70
bike-share: in Johannesburg 8, 22, 33–34, 36
Bissell, D. 105
Black and Minority Ethnic (BME) 24, 40
Black Chicago: The Making of Negro Ghetto (Spear) 41
Black geographies/transportation 12, 14, 18, 38–39; from Anglophone geography 39, 42; for automobilities 44; community action and protest 49; critical reflections 38; empirical and theoretical engagement with 41; informal 50–51; informality and dollar vans 46–47; in Johannesburg

47–51; labor exploitation and surplus value 43; legal systems 45–49; multiple consciousness 47; protesting transportation racism 49; public transportation 44–45, 48; racial immobility *vs.* mobility 45; scholarship 38, 41, 44, 52; spatiality 43–44; tactics for 45–47, 51–52; Whiteness of geography 39–41
The Black Metropolis (Drake) 41
Bogota 31–33, 121
Bottone, E. 44
Brand, B. 103; *Masstransiscope* 103
Bree Street Taxi Rank *110*, 111
Bremner, L. 74, 75
Brooks, A. 88
bubonic plague 116, 118–120
Buchanan, C. 70
Budd, Z. 111
Bullard, R. 45, 46
Buller, H. 86
buses: and apartheid spatial planning 5–6; Gauteng Household Travel Survey 7–8; public transportation failure 1, 5–6, 11, 15–16, 32, 35, 53, 60–62, 68, 76, 79, 112, 130, 133, 134
bus-oriented development *see* people-oriented transportation (POT)
bus rapid transit (BRT) 17, 22, 28, 30; adoption 58; failures 6–7; global proliferation of 56; Klipfontein Corridor, Cape Town 59–61, 64; minibus taxi industry 50; MyCiTi, Cape Town 7, 60, 131; policy mobilities 58; postapartheid 31; Rea Vaya bus 32, 61, 62; from taxi ranks 107
Bus Stops (Gomi) 127
Byrne, D. 96

Calthorpe, P. 54, 56; *The Next American Metropolis* 54
capability approach 12–13
Cape Town 1, 6, 7, 9, 12, 16, 17, 19, 20, 30, 31, 33, 36, 53–55, 58–61, 65, 70, 72, 89, 93, 96, 97, 100–102, 104, 107, 109, 111–113, 118, 119, 123, 124, 130, 131; art on transportation 107, 109; Bicycle Empowerment Network in 70; BRT systems 31; electric trams in 30; Foreshore Freeway 131; horse-drawn trams in 30; Klipfontein Corridor in 59–61; metropolitan municipalities in 1; Mobility Strategy 59; MyCiTi BRT 1, 6, 7, 60, 61, 65, 107, 109, 110, 131; people-oriented transportation (POT) in 58–61; transit-oriented development 58; transportation in 9
captive walkers 8, 67–68
Carter, D. 113
Casper the Commuting Cat (Finden) 89
Cayton, H.R. 41
Cervero, R. 55
Césaire, A. 42
Chancellor House Station Eastbound 108
Chang, I.-C.C. 2
Chatman, D.G. 56
Chicago School of Urban Sociology 88
Chinodya, S. 95
Cissokho, S. 11
city planning 15, 73, 88
Clarke, W.E. 85
Clayton, D.W. 23
Cole, E. 113
colonialism: internal *vs.* external 25; knowledge systems 25; of modernity/rationality 23
coloniality-induced institutional racism 24, 40
Connolly, C. 117
content analysis 16–17
Cooke, S. 55
Corridors of Freedom 61–64, *62*, 76–77, 128, 129
corruption complex 11
cosmopolitanism 36
COVID-19 pandemic 51, 97, 115, 119–125; restricting international transportation 123–124; restricting mobility 122–123; restricting urban public transportation 124–125
Crenshaw, K. 42
Cresswell, T. 45
critical art geographies 99; *see also* transp-art-ation
critical race theory 42
critical visual methodology 98, 104
Crot, L. 58
Crow, J. 89, 93
Crush, J. 26
cultural artifacts 86, 91
cultural geographies 40, 86, 99, 104; animal 86
Cycle Bridge 77, *79*
cycling 8, 12, 121; Johannesburg 33, *34*

Daily Maverick 16, 17
Dar es Salaam 11, 31, 36, 72
decolonizing geographies/transportation 3, 10, 12, 18, 82; during apartheid 30–31, 35–36; approaches 22, 24; bike-sharing

scheme 34, 36; capitalist-colonialist conditions 24; challenges 27–28; defiant scholarship 26; disciplinary reflection 22; horse-drawn trams 30; indigenous transport 29; informal *vs.* formal 29; internal *vs.* external colonialism 25; knowledge 14, 21, 23; methodologies 25; minibus taxis industry 22, 35–37; pro-poor agenda 30–32, 34–37; refining local knowledge 35–37; reinterpreting knowledge 31–33; rejecting external knowledge 33–34; scholarship 22, 24; social and spatial transformation 36; tactics for 30, 31
DeKalb Avenue 103
Demonic Grounds (McKittrick) 43
Department of Small Business Development 123
Desai, V. 24, 40
Deskins, D. 39, 40
developer-propagated mobility ideologies 105
development-oriented transit *see* transit-oriented development (TOD)
Didier, S. 57
Drake, St. Clair 41
Driver, F. 22
Du Bois, W.E.B. 47, 51; *The Philadelphia Negro* 41
Duffy, R. 84, 90
Dugard, J. 36
Durban 1, 30, 91, 100, 114, 118, 119, 122; *see also* eThekwini
Dutch East India Company 93
Dwyer, O.J. 39–40

Ebola 117, 120–121, 123, 126
Economic Freedom Front (EFF) 61, 132
electric trams 30, 54, 71, 74, 93; *see also* trams
Ellis Park 107
Emel, J. 86
empire education 25
Enright, T. 103
epistemic knowledge 21
epistemological decolonization 23
Espi, L. 109
Esson, J. 24, 40
Essop, T. 59, 60
eThekwini 1, 5, 6, 17, 55, 114, 123
eThekwini Metropolitan Municipality 114
ethnography of transportation workers 11
ethology 91
European colonization 22–23
Ewing, R. 117, 122
extrospective transfer 2

failure *see* transportation failure
#fallism 27
Fanon, F. 22, 42
#FeesMustFall 27, 102
Finden, S.: *Casper the Commuting Cat* 89
Flack, A.J.P. 89
Florida, R. 117, 118, 126
Forde, S. 11
formal transportation 15; decolonizing geographies/transportation 29; *see also* transportation
Forsyth, A. 73, 74
French geographers 23–24
Fuller, G. 105

garden city model 58, 59
Gauteng 4, 5, 7, 8, 64, 68, 125
Gauteng Household Travel Survey 7–8
Gautrain, rail transportation service 4–5, 124; Sandton Station *5*
Gaylard, G. 82
geographies 2, 3, 10, 14; cultural 40, 86, 99, 104; health 10, 13, 15; hidden 27; historical 12, 40; interdisciplinary 10, 15; peoples 26–27; transport 2, 3, 10, 12–15, 17, 18, 22, 27, 29, 44, 47, 51, 52, 65, 85, 91, 97, 99, 100, 103, 112, 114, 128, *see also specific geographies*
Gibbs, L. 86
Gilmore, R.W. 43
Glass, T. 46
global south 10, 56, 57, 85, 90, 117, 121; city planning in 47, 82, 117; and transport in 14, 16, 19, 27–29, 31, 47, 51, 69, 71, 82, 98, 107, 114, 127, 132
Global Walkability Index 69
Go Durban 1, 6
Golden Arrow Bus Services (GABS) 1, 5
Gomi, T. 127; *Bus Stops* 127
Gontsana, M.-A. 124
González, S. 57
Gormley, A. 104
Gould, P.R. 28
government-built housing 131
Grayston Pedestrian Bridge 77–80, *79*
Great Walk 77
Green Book 45
Growth and Development Strategy (2040) 76
Guinard, P. 102
Gurney, K. 102
Gwala, S. 108

Hamidi, S. 117, 122
Harris, A. 57, 58

Harrison, P. 62
Hart, G.H.T. 74
Harvey, D. 40
Hawkins, H. 104
Hawthorne, C. 38, 43
Hayhurst, L. 11
health geographies 10, 13, 15; *see also* geographies
Healthy Cities Project (1978) 117
Herron, B. 109, 111
hidden geographies 27; *see also* geographies
hijacked buildings 130–131
Hinger, B. 44
historical geographies 12, 40; *see also* geographies
Hodgetts, T. 90, 91
Holloway, L. 92
Hook, W. 7
horse-drawn trams 9, 27, 71, 93; decolonizing geographies/transportation 30; walking-as-transportation/walkability 74; *see also* trams
Hovorka, A. 84, 86, 87, 89
How Cities Learn (Wood) 17
How the Other Half Lives (Riis) 116
Hubbard, P. 88
human geography: animals in 85–87; *see also* animal geographies
Human Rights Wall 100

imperialism 22, 26
inclusionary housing policy 62–63
India 120, 121
infectious disease 117, 119, 121
informal transportation 8, 11, 12, 14–16, 29, 31, 36, 45, 50, 52, 63, 103, 106, 129, 132; Black geographies 45, 50–51; decolonizing geographies/transportation 29
Institute for Transportation and Development Policy 69
interdisciplinary geography 10, 15; *see also* geographies
international travel 123–124
introspective failure 2
Inwood, J. 45
Isnard, H. 24

Jacobsen, M. 11
Jacobs, J. 70, 114
Jamme, H.-T. 55, 56
Jazeel, T. 21, 25, 37
Joburg Photo Umbrella 112
Johannesburg: animal labor 93; bike-share in 33–34; Black transportation in 47–51; Bree Street Taxi Rank *110*, 111; Corridors of Freedom 61–64, *62*, 76–77; cycling in 33, *34*; horse-drawn tram 9; Louis Botha Avenue 61, *62*, 106, *106*; metrobus *6*, *64*; minibus taxi in *35*, 35–36; Non-Motorized Transport Framework 76; postapartheid transformation in 34; Public Art Policy (2006) 101; racial and gender equality 9; Rea Vaya *see* Rea Vaya, Johannesburg; Shosholoza Meyl 3–4, *4*; walking-as-transportation/walkability 19, 67–84
Johannesburg Development Agency (JDA) 16, 32, 77, 80, 81, 101, 107

Kant, I. 39
Kaur, T. 11
Kębłowski, W. 21, 29, 30, 35
Keil, R. 117, 120
Kendall, C. 117
Kentridge, W. 101
Khosa, M. 12, 27
Klipfontein Road, Cape Town 59–61, 64
knowledge: colonialism 25; decolonizing geographies/transportation 14, 21, 23; epistemic 21; refining local 35–37; reinterpreting 31–33; rejecting external 33–34
Kobayashi, A. 39
Kockelman, K. 55
Krambeck, H.V. 69
Kruger, P. 100
KwaThema, hand signal for 113

Lagos 11, 31, 36, 117, 120
Lauermann, J. 2
Leach, M. 117
Leduc, G. 24, 25
Lefebvre, H. 13
leisure cyclists 33
Liebow, E. 41–42; *Tally's Corner* 41–42
Lipsitz, G. 43
Litman, T. 69, 70
Lomp, S. 127; *Things that Go* 127
London 2, 16, 17, 23, 54, 87, 103, 116, 120, 121, 133
Lo, R.H. 73, 74
Lorimer, J. 90, 91

Louis Botha Avenue: Johannesburg 61, *62*, 106, *106*
Lovell, H. 2
Lucas, K. 80
Lunstrum, E. 84, 90
Luthuli, W. 101; *Angel of the North* 101

Manila 16
Market Photo Workshop 112
Martin, B. 70
Marx, G. 101
Marxist geographers 40
Mashaba, H. 61
Massachusetts Bay Transportation Authority 103
Massé, F. 84, 90
Masstransiscope (Brand) 103
Matlon, J. 38
Maxwele, C. 102
Maylam, P. 119
Mbalula, F. 125
McCann, B. 70
McCaul, C. 36
McGregor, J. 84
McKittrick, K. 41, 43–44, 51; *Demonic Grounds* 43–44
metaphorical significance of syndrome 119
Metrobus 5, *6*; bus stop, Johannesburg *64*
metropolitan municipalities 1
Metrorail 1, 4, 7
Middleton, J. 29
Mignolo, W. 21, 23, 29–31
Milpark Pedestrian Bridge 80–81
minibus taxi industry 1, 8–9, 22, 128; Black geographies 50; bus rapid transit (BRT) 50; decolonizing transportation 22, 35–37; hand signals 113; in Johannesburg *35*, 35–36; mobile art 111; pandemic effects 124
Minty, Z. 101
mobile art 111–112
mobility justice 13
Mobius, C. 85
Mohtar, A. 29
Montgomery bus boycott 46
Moore, L. 84, 90
Moore, S. 57, 58
Moosajee, R. 31, 107
moral masturbation 40
Morange, M. 57
Morgan, N. 12
Morrill, R.L. 28
Motor Carrier Transportation Act 9
Munger, E. 23
Murray, L. 99
MyCiTi BRT 1, 6, 7, 60, 61, 65, 107, 109, *109*, 110, 131

Naidoo, S. 112
Nairobi 16, 36, 72, 119
National Department of Transport 3
National Taxi Alliance (NTA) 50
Native (Urban Areas) Act 1923 118
Negritude Movement 24, 42
Neto, K. 111
Newbigin, M. 85
Newcastle 133
New Deal 104
Newtown Cultural Precinct 101
The Next American Metropolis (Calthorpe) 54
Njoh, A.J. 28
non-motorized transportation 1, 12, 14, 72, 77; in Gauteng 8, *8*; on Grayston Drive 79; modes 133; *see also* transportation
Nyovest, C. 114

Orlando Police Station 108
OR Tambo International Airport 4
Oviedo, D. 72

Parker, A. 12
Parkins, A. 39
Parks, R. 45, 46
Passenger Rail Agency of South Africa (PRASA) 3, 125
Peck, J. 57
pedestrian 73, 74; bridges 77–79; Grayston Pedestrian Bridge 77–80, *79*; Milpark Pedestrian Bridge 80–81; profile and practices of 76; projects variety 77; types of 73; walk in blocked cycle lanes *75*, 75–76; Westbury Pedestrian Bridge and Park 81–82; *see also* walking-as-transportation/walkability
Pedestrians First: A Tool for Walkable Cities 69
People for the Ethical Treatment of Animals (PETA) 86
People Mover (eThekwini) 1, 5
people-oriented transportation (POT) 53; in Cape Town 58–61; in Johannesburg 58, 61–63, *62*; *see also* transit-oriented development (TOD)
peoples geography 26–27; *see also* geographies

Performing Animals Protection Act (1935) 94
Petersen, T. 112
Peyroux, E. 57
Phadi, M. 47
The Philadelphia Negro (Du Bois) 41
Phillips, H. 118–119
Philo, C. 86, 88
Pierre, J. 38
Pirie, G. 12, 27
planning: apartheid spatial 2, 9, 31, 32; colorblind transit 45; informal transportation 50–51; policy actors 2; South African cities 19; transportation 7, 9, 13, 14, 17, 19, 63–65, 68, 70, 82, 97, 131; urban 2, 26, 47, 53, 71, 72, 81, 117, 118
Plessy, H. 45
policy failure 2, 19, 57, 58, 65
policy maker 3, 11, 30, 72, 100, 101, 114, 132
policy mobilities 3, 10, 11, 13, 14; failure 19, 58; people-oriented transportation 53; scholarship 65; transit-oriented development (TOD) 57–58
policy taker 3
political ecology 87
Port Elizabeth 30
Porter, G. 11
Portrait with Keys (Vladislavić) 82
Pretoria 12, 27, 30, 49, 100
Prevention of Cruelty Act (1914) 94
Prince, R. 58
private transportation 1; *see also* public transportation
Progress of Human Geography 86
pro-poor agenda: decolonizing geographies/transportation 30–32, 34–37; walkability 81
public art 99, 100; culture-led regeneration 114; installation of 107; policy for 104; political and economic reasons 101; postapartheid 100–101
Public Art Policy (2006) 101
public health 116–118
Public Health Act 118
public transportation: Black geographies 44–45, 48; declining bus usage 7; dog-friendly 87; failure repeats for buses 5–6; formal 99; horse-drawn tram 9; informal 35–36, 99; labor exploitation in 10; modes of 124; motorized forms of 68; responsibilities 3; subsidies 9; supply and demand for 124; urban 121–122, 124–125
Public Utility Transport Corporation (PUTCO) 1, 5, 49

Queen Elizabeth Bridge 101
Quijano, A. 21, 23, 30, 31

racial capitalism 38, 40
rail transportation service 3–4; Gautrain 4, *5*; high-speed inter-urban 5; Shosholoza Meyl 3–4, *4*
Rand Daily Mail 16, 17, 118
Rawls, J. 13
Rea Vaya, Johannesburg 6, *7*, 17, 127; articulated and bi-articulated buses 32; bus rapid transit (BRT) 32, 61, 62; glass etching 107, *108*; on Louis Botha Avenue 106, *106*; seats on bus *32*, 32–33
research positionality 17
Rhodes, C. 100, 102
#RhodesMustFall 27, 102
Ridgeback, R. 95
Riebeeck, J. van 93
Riis, J. 116; *How the Other Half Lives* 116
Rink, B. 12, 85, 89, 93
Rizzo, M. 11
Robinson, C. 40
Robinson, J. 57, 62
Rodney, W. 25
Rodríguez-Pose, A. 117, 118, 126
Rose, G. 98, 104
Rose, H. 40, 42
Royal Geographical Society-Institute of British Geographers (RGS-IBG) 21
Rubin, M. 12
rural *vs*. urban transportation 1

Sabouri, S. 117, 122
Said, E.W. 22
Samaha, P. 29
Sandton (Gautrain Station) 4, *5*
sanitation syndrome 118
San rock art 100
Santos, M. 25–27
Sao Paolo subway station 98
SA Taxi Foundation Art Award 111
Schwanen, T. 12, 13, 21, 28–29
Scott, W. 45
segregated bus system 9
Semenya, C. 111

Sen, A. 12
Severe Acute Respiratory Syndrome (SARS) 117, 120
Shabazz, R. 43
Sheller, M. 12, 13
Shields Ferry 133
Shingne, M.C. 88
Shosholoza Meyl: in Johannesburg 3–4, *4*
Silk Road 119, 120
Simelane, B.C. 125
smallpox 118–120
Smith, N. 40
Soja, E.W. 13
South Africa: art in postapartheid 100–102; Blackness 47; BRT 31–33; COVID-19 impacts on transportation 122–125; decolonizing geography in 25–27; decolonizing transportation in 30–37; informal public transportation system 35; people-oriented transportation 53–65; transp-art-ation in 106–114; trans-paw-tation" in 84, 85, 92–96; transportation in 1, 3–10; transportation planning 63–65; urban mobilities scholarship 12
South African cities 1, 3, 7, 15–19, 30–32, 50, 53, 57, 64, 66, 74, 84, 85, 92, 98, 100, 115, 116, 124, 127, 132
South African Cities Network 17, 50
South African government 9, 27, 123
South African history 9, 12, 25–26
South African National Taxi Council (SANTACO) 50
South African transport 6, 12, 30, 38
Southeastern Geographer journal 41, 42
Southworth, M. 68
Soweto Uprising 26
Spanish flu 118, 119
spatial justice 13, 14, 70
Spear, A. 41, 42; *Black Chicago: The Making of Negro Ghetto* 41
Spencer Commission of Enquiry 48, 74
Stasik, M. 11
Stellenbosch University 25
Stent, J. 124
Storper, M. 117, 118, 126
Strategic Integrated Transport Framework 76
Suremain, M.-A. de 23
Swanson, M.W. 118
Swart, S. 90, 95
Swartz, N.D. 113
Taaffe, E.J. 28
Tally's Corner (Liebow) 41–42
Tau, P. 61
Taxi Recapitalization Program 50
Taxi Relief Fund (TRF) 51, 125
taxi wars 36
Temenos, C. 2
That's Not My Bus (Watt) 127
Theodore, N. 57
Things that Go (Lomp) 127
Thiong'o, N. wa 26, 27, 30, 33
Thokoza Park 107
town planning 2, 59
trams 27, 48, 54, 74; electric trams 30, 54, 71, 74, 93; horse-drawn trams 9, 27, 30, 71, 74, 93
trains 1, 4, 10, 11, 17, 27, 35, 36, 89, 99, 103, 107, 112, 113, 119, 122, 124, 125
transit-oriented development (TOD) 53–54; application 55; best-practice policy models 57; as financial tool 55; garden city model 58; label 55; limitations 63–64; macro-and micro-transactions 64; and non-motorized mobilities 56; people-focused development approach 56; phrasing of 63; policy mobilities 57–58; principles 55; problematizing 54–56; scholarship on 56, 65; *see also* people-oriented transportation (POT)
transp-art-ation 19–20, 98–114; analysis 98, 105; apartheid and postapartheid era 100–102; critical approaches to 99–100; critical visual methodology 98; critique of 105; deterritorializes geography 99; inclusive approach 114; informal and formal characteristics 105–106; installations 111; mobile art 111–112; and mobility 99; on MyCiTi 109; representations of 112–114; spatializing of 104; state and local art programs 103; video and visual arts 114; visualization sites 98
trans-paw-tation 19, 84–97; animal geography 85–87; Animal Protection Act No. 71 of 1962 94; biopolitics of animal control 91; epistemological and methodological approaches 90–91; features 92; imagery and aesthetics of animal mobilities 95–96; movement of animals 92–94; policies governing animal mobilities 94–95; Prevention of Cruelty Act (1914) 94; in South African

cities 92–96; transportation failure 84, 92, 96, 97; urban paws 87–89
transportation: aesthetics on 98; animal geographies and 89–92; art roles in 15, 103–111; Black 38–52; BRT.6, 7; and COVID-19 119–125; decolonization 10, 18, 21–37; failure 115–126; geography 14, 17, 22, 27; governmental responsibilities for 3; horse-drawn 93; informal 8, 11, 16, 36, 50–51; infrastructure 28; justice 3, 12–15; mobilities and 21, 119; motorized 13, 14, 33; non-motorized 1, 8, *8*, 14; people-oriented 53–66; planning 7, 9, 13, 14; policy mobilities and 57–58; public 1, 3, 7, 9, 16; racial and gender equality in 9; reflecting on 130–134; representations of 112–114; restricting international 123–124; in restricting urban 124–125; South Africa 3–10, 130–132; urban 1, 3, 9, 11; sustainable 4; walking-as 67–83; workers in Dar es Salaam 11; *see also individual entries*
transportation failure 92, 114–126; animal mobilities 84, 92, 96, 97; Black transportation 38–52; bus rapid transit (BRT) 6–7, 32; colonial transportation systems 29; concept 2; during COVID-19 pandemic 119–125; introspective 2; methodologies of 15–18; policy 2, 57; policy mobilities 58; public transport 35; reflecting on 127–130; restricting international transportation 123–124; restrictions on mobility and 122–123
transport poverty 28
Trubina, E. 29
Tsele, M. 113
Tshwane 1, 4–7, 16, 17, 31, 33, 97, 107, 124
Tuck, E. 25, 37
Tuvikene, T. 21, 29, 30, 35
Twentieth International Geographical Union 23

Universal Declaration of Human Rights 100
University of Johannesburg 12, 25, 108
urban geography 21, 40, 88; animal 87–89; *see also* animal geographies
Urbanik, J. 86
urban transportation 116–118; of Ebola and SARS 117; financing and development of 9; mobilities 3, 10–12; modes 11; rural *vs.* 1; urban public transportation 121–122, 124–125
US Department of Transportation 69

Van Sittert, L. 90, 95
vector-borne diseases 117
Venter, C. 12
Verlinghieri, E. 12, 13, 29
Verne, J. 103
visual analysis 17
Vladislavić, I. 82; *Portrait with Keys* 82

walking-as-transportation/walkability 19, 67–84; apartheid and postapartheid era 74–75; challenges 72, 82; complete streets policy 70–71, 77; Corridors of Freedom 76–77; difficulties of 73; economic and social benefits 71, 73; European cities 71; features 69–70; horse-drawn trams 74; in Johannesburg 74–82; lack of capacity/funding 72; limitations 72; literature on 72–73; Milpark Pedestrian Bridge 80–81; multidisciplinary approach 68; physical factors 69; planning for 68, 70; policy 68, 72; qualitative analysis of 69; reconceptualizing 68–71; road diets 70; routine accommodation 70; scholarship 71; and street-level interactions 76; terminology 70; universal design 70; unpleasantness of 80–81; urban design and 71–72; Westbury Pedestrian Bridge and Park 81–82; *see also* pedestrian
Walters, C. 78
Walters, J. 12
Ward, K. 2
Washinyira, T. 124
Watt, F.: *That's Not My Bus* 127
Weinstock, A. 7
Welford, M. 120
Wells, H.G. 103
Wells, K. 2
Westbury Pedestrian Bridge and Park 81–82
Western Cape 4–6, 59, 125, 130
Whiteness of geography 39–41
White Paper on National Transport Policy (1982) 9, 36
Whol, M. 70
Wilbert, C. 86, 88
Wilkinson, A. 117

Wilkinson, K. 92
Williams, K. 44
Williams, M. 59
Wilson, B. 43
Wilson, R.T. 85
Wohlwill, D.E. 56
Wolch, J. 84–86, 88
Women's Political Council 46
Wood, A. 11, 21, 29, 30, 35, 57; *How Cities Learn* 17
Woods, C. 41, 43
Woolf, S. 113
Wright, H.K. 47

Yang, K.W. 25, 37
Yong, E. 123

Zavetoski, S. 70, 71
zoonotic diseases 97, 117
Zuidgeest, M. 55
Zuma, J. 95
#ZumaMustFall 27

FSC
www.fsc.org
MIX
Papier | Fördert
gute Waldnutzung
FSC® C083411

Kolibri 360 GmbH
Ferdinand-Jühlke-Straße 7
99095 Erfurt, Deutschland
produktsicherheit@kolibri360.de